Nature, Business, and Community in North Carolina's Green Swamp

UNIVERSITY PRESS OF FLORIDA

Florida A&M University, Tallahassee
Florida Atlantic University, Boca Raton
Florida Gulf Coast University, Ft. Myers
Florida International University, Miami
Florida State University, Tallahassee
New College of Florida, Sarasota
University of Central Florida, Orlando
University of Florida, Gainesville
University of North Florida, Jacksonville
University of South Florida, Tampa
University of West Florida, Pensacola

Nature, Business, and Community in North Carolina's Green Swamp

Tycho de Boer

University Press of Florida
Gainesville/Tallahassee/Tampa/Boca Raton
Pensacola/Orlando/Miami/Jacksonville/Ft. Myers/Sarasota

Printed in the United States of America on recycled, acid-free paper

13 12 11 10 09 08 6 5 4 3 2 1

Library of Congress Cataloging-in-Publication Data
Boer, Tycho de.
Nature, business, and community in North Carolina's Green Swamp / Tycho de Boer.
p. cm.
Includes bibliographical references and index.
ISBN 978-0-8130-3248-1 (alk. paper)
1. Green Swamp (Brunswick County and Columbus County, N.C.)—Environmental conditions. 2. Green Swamp (Brunswick County and Columbus County, N.C.)—Economic conditions. 3. Forests and forestry—North Carolina—Green Swamp (Brunswick County and Columbus County)—History. 4. Capitalism—Environmental aspects—North Carolina—Green Swamp (Brunswick County and Columbus County)—History. 5. Longleaf pine—Economic aspects—North Carolina—Green Swamp (Brunswick County and Columbus County)—History. 6. Land use—North Carolina—Green Swamp (Brunswick County and Columbus County)—History. 7. Nature—Effect of human beings on—North Carolina—Green Swamp (Brunswick County and Columbus County)—History. 8. Human ecology—North Carolina—Green Swamp (Brunswick County and Columbus County)—History. 9. Community life—North Carolina—Green Swamp (Brunswick County and Columbus County)—History. 10. Green Swamp (Brunswick County and Columbus County, N.C.)—Social conditions. I. Title.
GE155.G744B64 2008
333.91'809756—dc22 2008014482

The University Press of Florida is the scholarly publishing agency for the State University System of Florida, comprising Florida A&M University, Florida Atlantic University, Florida Gulf Coast University, Florida International University, Florida State University, New College of Florida, University of Central Florida, University of Florida, University of North Florida, University of South Florida, and University of West Florida.

University Press of Florida
15 Northwest 15th Street
Gainesville, FL 32611-2079
http://www.upf.com

Caleb was in a difficulty known to any person attempting in dark times and unassisted by miracle to reason with rustics who are in possession of an undeniable truth which they know through a hard process of feeling, and can let it fall like a giant's club on your neatly-carved argument for a social benefit which they do not feel.

George Eliot, *Middlemarch*

Contents

Figures

Acknowledgments

The completion of this book comes at the end not only of the usual academic trials and tribulations but also of a dark and difficult period in my personal life. It seems fitting, therefore, to thank first those people who stood by me through what was truly a vale of tears. Without the unwavering love and support from my wife Melissa and our children Zoe and Elliott (the latter being, to be perfectly honest, helpfully oblivious to all of it), this work would not have seen the light of day. Also my parents, Christiaan and Dory de Boer; Pastor Joe Weaver and his "posse" at the Fairall Charge; and my dear friend Angela Smith—all were indispensable shoulders to lean on and inexhaustible sources of wonderful help and support. I also must thank my sister Carmen Mutsaers and her family; my in-laws, Rich and Jan McCabe; Kathy Roberts; Virgil and Bobbie Boelter, for providing me with a quiet writing space during the summer of 2006; my friends Andrew McMichael, Al Granath, and Tim Weber and their families for their hospitality during research trips; Chuck Lee, Kendall Staggs, Mariana Candido, Shelley Sinclair, and the rest of the history department at the University of Wisconsin–La Crosse for an intellectually, emotionally, and financially supportive environment; Daniel Adamski, John Thompson, Frank Pazzynski, and Paul Sielski at Waynesburg College for keeping me on track; and Mathieu Klapwijk, Neal Roellich, and Nick Gaynor for helping me create a musical outlet for my anxieties and frustrations.

At the same time, I would not have been able to finish this book without the generous financial support of Vanderbilt University, Waynesburg College, and the University of Wisconsin–La Crosse, nor without the research grants I received from the North Caroliniana Society at the University of North Carolina–Chapel Hill, and the Forest History Society in Durham, North Carolina. The helpfulness and expertise of the staff at these two societies, as well as of the staff at the Southern Historical Collection at the University of North Carolina–Chapel Hill, the Rare Book, Manuscript, and Special Collections Department of the Perkins Library at Duke University, the North Carolina State Archives in Raleigh, the Special Collections Department of the J. Y. Joyner Library at East Carolina University, the Special Collections Department of the William Madison Randall Library at the University of North Carolina–Wilmington, the Special Collections Department of the Z. Smith Reynolds Library at Wake Forest University, and the Historical Collections Department of the Baker Library at Harvard Business School made my yoke a lot easier, my burden lighter, and at times quite joyful to bear. I am also forever indebted to Meredith Morris-Babb, Eli Bortz, Michele Fiyak-Burkley, and the rest of the staff at the University Press of Florida, whose patience and faith over the years have been near-saintly. Further encouragement and crucial insight came from Wil Verhoeven and Edward Ayers, who together may be held responsible for turning me on to Southern history, obscure literature, and the trimmings and trappings of academia; from Jack Temple Kirby, Stephen Taylor, and Randall Miller, all of whom read several drafts of the manuscript; and from Stacey Robertson, Otis Graham, Louis Kyriakoudes, Michael Thompson, and Christine Meisner Rosen. At Vanderbilt University, the many suggestions and criticisms proffered by Andrew McMichael, Ed Harcourt, Anthony Stanonis, Mike Crane, David Karr, Michael Bess, Don Doyle, Joyce Chaplin, James Epstein, Cliff Russell, Jane Landers, and Helmut Smith during seminars and private conversations also proved extremely helpful long after I left that wonderful place.

Finally, I must express my gratitude to two people who guided me through both my personal and my academic travails. First, David Carlton steered me toward the longleaf pine forest as a research subject, sharpened my thinking on the role of business in American and Southern history, introduced me to the wonderful communal experience of shape-note singing, encouraged me to read James C. Scott, urged me to approach the articulation of my arguments as writing a "hook" for a country song at a time when my writing reflected my appreciation for Mahler's symphonies, and joined me for quite a few uplifting—or mind-erasing—cups of coffee at Bongo Java in Nashville. I am proud to have worked with him and even more delighted that I may call him my colleague and friend. Second, Scott Williams has been not only a steadfast friend (or "buddy," as he prefers) who shared with me the goodness of Equal Exchange coffee, he also introduced me to the writings of Wendell Berry, John Howard Yoder, Alasdair McIntyre, William Cavanaugh, and Karl Barth, by way of which he taught me to challenge academic orthodoxy, question "common sense" as "common to whom?," perceive of hunting as anything but "murdering Bambi," and think of particular understandings of history as part of comprehensive worldviews, of stories and meta-narratives about the world, and the purpose of human existence. I hope this book reflects David's and Scott's influence, and it is to them that I dedicate this book.

Introduction

"The Magic Wand of Capital and Enterprise"

In 1992 the Federal Paper Board Company of Riegelwood, North Carolina, took out a full-page advertisement in the *Whiteville News Reporter* to announce the delivery of three oak trees to every school campus in Columbus County, a largely rural county in southeastern North Carolina. Adorning the advertisement with a picture of Christopher Columbus' landing in the Americas five centuries earlier, the company billed its initiative as the kind of "exceptional idea" that only comes along "every 500 years or so." Dwelling on its own history and place in Columbus County, the manufacturer of pulp and paperboard products proudly claimed to have "enjoyed being an economic force in Columbus County since 1951," and to have "contributed over $254.5 million to the local economy" in 1991 alone. It conceived of its gift to local schools of a trio of oak trees as a way "to do more for our community," and, after professing its belief "in the reforestation of our land," expressed the hope "that teachers will use this event to focus on the importance of ecology and recycling." After all, the ad concluded, "At Federal Paper, we believe in contributing to our community, and children are our favorite natural resource."[1]

Of course, whether "our land" referred to any land the people of Columbus County might call "theirs" as a community or to the company's more than 400,000 acres of corporately owned timberland remained unclear. It is easy to dismiss such ads as so much self-aggrandizement and corporate hubris about care for land and community: the pulp and paper companies on the coast of southeastern North Carolina consistently rank among the region's worst water and air polluters, a dubious honor they share with the booming hog industry and with lumber-processing companies like Fayetteville's Cape Fear Wood Preserving, whose plant is a federal Superfund site. In 1996 the Environmental Working Group listed the Riegelwood plant, by then part of International Paper after a $3.4 billion merger,

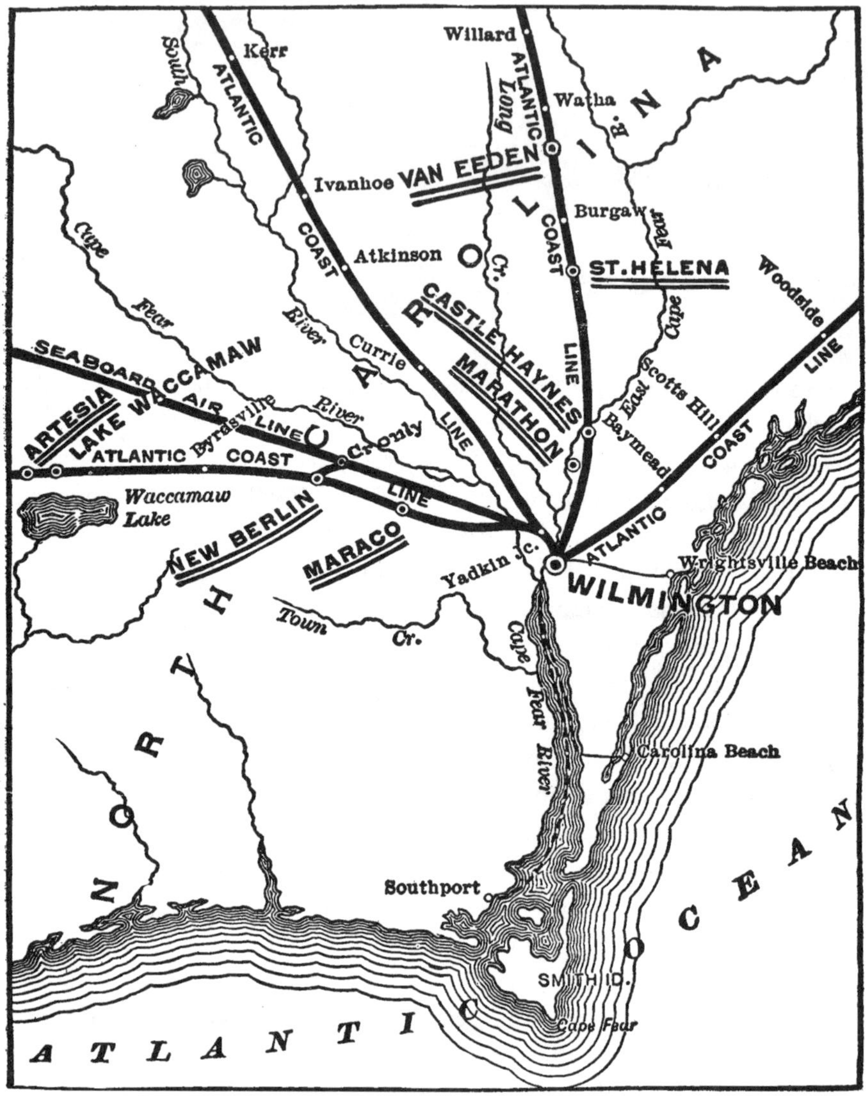

Figure 1. Map of southeastern North Carolina from the promotional literature of the Carolina Trucking Development Company. The map reflects well the perceived "emptiness" of the Green Swamp, which is located to the southeast of "Waccamaw Lake." (Courtesy of Hugh MacRae Papers, Duke University.)

as one of the three worst particulate polluters in North Carolina.[2] The modern industrial incarnations of what have historically been two of the Lower Cape Fear's staple economic activities—forest industries and hog farming—have come under increased fire from environmentalist groups who decry the ongoing transformation of one of the United States' most ecologically diverse areas at the hands of large corporations. Organizations such as Friends of the Green Swamp, Rainforest Action Network, the North Carolina Environmental Defense Fund, and Swamp Watch Action Team have called alarmed attention to the devastating impact of large-scale pig farming, mono-crop tree farming, and industrial pulp manufacturing on the area's wetlands. Of special concern to these organizations is the growing threat industry poses to the 15,000 acres of land and water that comprise the Green Swamp in Brunswick County, east of Columbus: despite management by the Nature Conservancy, this swamp and its environs continue to inspire plans for prospective development.

Moreover, the exploitation and devastation, at the hands of enterprising capitalists, of the wetland ecosystems in and around the Green Swamp provoked voices of protest long before modern environmentalists prophesied the wetlands' likely disappearance in the face of the ever-expanding operations of large corporations. In 1905 Walter Gilman Curtis, writing about the history of Wilmington and Southport, lamented the irreversible destruction the various forest industries had wrought in the expansive swamps and longleaf pine forests located along the rivers to the west of those towns. The land, he wrote, used to be "covered with the most beautiful growth of pine that it was possible to imagine," but once the naval stores industry had its way with these trees, everything had changed:

> If one now . . . goes out into these lands to examine them, he will find they have been hacked as high as the laborer can reach . . . and the turpentine exhausted in the tree never to be replaced; that there [*sic*] beauty is gone, and that . . . very frequently fire has gone through these inflammable trees and destroyed them outright and forever.

If "outright and forever" destruction was not enough, matters had only gotten worse after the decline of turpentine extraction:

> At this point the lumber getter makes his appearance . . . Pines, cypresses, black gum and every other tree which can be shipped and manufactured into anything is cut and prepared for shipment . . . Thus

> the clearing of North Carolina of every useful trade goes on with fearful rapidity and the owners of the lands find themselves but very little richer . . . and the glory and beauty of the scenery is destroyed forever.

The forest industry, thanks to the complicity of short-sighted local landowners, had turned trees into money, the longleaf pine belt into a wasteland, and its back on the area's further development—a typical tale of invasive capitalist greed, ill-advised local compliance, and nature attacked and squandered.[3]

Modern environmentalists have certainly made extensive use of this template to explain the ecological degeneration and destruction around the Green Swamp and elsewhere, the scars of which cover so much of the American landscape. Janisse Ray is one of several recent spokespersons for "the pain of the lost forest"—environmentalists who have given voice to a spreading perception of commercial forest exploitation as inimical to an environment that transcends human comprehension, action, and design. Put bluntly, "God doesn't like a clearcut," while newly planted "pine plantations dishearten" Him.[4] The late Warren Dean likewise wrote that "forest history . . . is everywhere on this planet one of exploitation and destruction."[5] The human population, he argued, was at the root of destructive environmental transformation across the globe: the larger a population in a given locality, the farther-reaching, and, inevitably, the more destructive the changes they wrought, making environmental degradation one of "the most consequential of human activities." Dean shared with many scholars and environmental activists the view that of all human endeavors, economic activity had the most invasive and lasting impact on the natural world; and since in the Americas capitalism had driven most economic activity, he called on fellow historians to dispel that ideology's "hollow triumphalism" regarding economic progress that so often masked the grim history of environmental degradation.[6]

Yet while such harsh words continue to resonate with those who hold capitalism, in the words of Gordon Wood, "ultimately responsible for much of what ails us,"[7] Dean did take care to nuance his overall indictment. Waste, recalcitrance, and sheer stupidity among American settlers had at times contributed more to forest destruction than the economic dictates of the imperial powers that sent them overseas. European potentates may have neglected "to regard the natural world as something more than a disposable collection of utilities," but in terms of "the effectiveness and efficiency with which scarce resources are employed . . . and the validity of the uses to which they are put," royal governments compiled a far better record than many of their subjects.[8] Capitalist exploitation, although very detrimental to environmental health and biodiversity, at least produced certain benefits—economic wealth, higher standards of living, and material goods—that wanton destruction did not.[9]

"The rugged woods are putting on city airs"

In other words, there is a different story to tell, a story of business development, community formation, and human-ecological interaction in and around the Green Swamp that is neither the environmentalist's declension narrative of destructive exploitation nor the corporation's triumphalist account of progressive and sustainable capitalist development.[10] There is, first of all, the irony that there would be little Green Swamp to defend had it not been for Federal Paper Board's donation of 13,850 acres of the current preserve to the Nature Conservancy in 1977 and of an additional 2,577 acres in the late 1980s. Second, a 1969 survey by the Southern Forest Experiment Station found that, in spite of three centuries of extracting turpentine, logging timber, and clearing land for agricultural purposes, forestland still covered more than half of the land in North Carolina's southeastern corner.[11] Biologists found, however, that covering much of the former longleaf pine belt was not a longleaf pine forest but a mixed hardwood forest. Forest users, inadvertently or purposely, had allowed parts of the forest to regenerate after years of extractive activity—activity at various levels of intensity and on various geographic scales—and as a result growths of mixed hardwoods and loblolly pine, as well as extensive pine tree farms, had sprung up amid cultivated lands to replace the original stands.[12] This mosaic of successive forest growths and remaining swamps, therefore, not only bore but also constituted a "human footprint" in the forest—the *presence* and *proliferation* of particular tree species revealed the human hand in nature as much as did the absence and disappearance of others.[13] While from an ecological perspective the disappearance of a specific, sometimes unique forest environment constituted a lamentable loss, from the perspective of local communities "the forest," "the woods," and "the swamp," in a cultural more than an ecological sense, continued to be a significant presence in their midst.

In this story forest industries were not altogether invasive, destructive, and transient, although there was plenty of that; like the forest itself, it was an enduring economic and cultural presence in communities on the edge of forest and swamp. In the same edition of the *News Reporter* that carried Federal Paper's advertisement, Council Tool of Lake Waccamaw—initially a manufacturer of turpentine tools, then of forestry equipment, and eventually of striking tools and drop forgings—lauded its own permanence as a stable economic force in Columbus County: "We've seen a lot of changes in the last 106 years," its ad read; "four wars, several recessions, one great depression, two stock market crashes and at least twelve men on the moon." In a similar ad four years later, the company added a destructive fire, "Republicans and Democrats in power, civil strife, and a multitude of blizzards, droughts, floods, tornadoes, and hurricanes" to the list of literal and figurative storms the company had weathered. Emphasizing its quality, consistency, and dependability in the face of the unpredictability of nature and the contingencies of human history, the company praised its local employees, many of whom had been "an important part of our company for decades," and some of whom were second generation workers at the plant.[14] Although this ad, too, had a certain chest-thumping quality to it, the claims it made were nonetheless by no means unfounded, and, as such, it forces us to acknowledge the ways in which forests and forest industries *can* coexist to produce shared benefits and communally shaped environments that meet the demands their human and non-human inhabitants place on it.

After all, until what Kirkpatrick Sale called the "Green Revolution" propelled environmentalism as a mass movement into American life and politics in the decades after World War II, voices such as Curtis,' Ray's, and Dean's were few, and usually those of prophets in the wilderness.[15] Even more, modern environmentalists would be wrong to assume automatically that an early critic like Curtis was "on their side," like them bemoaning for nature's sake the spoiling and destruction of the wilderness for the benefit of a greedy, capitalist, urban-industrial complex. Curtis was certainly lamenting the destructive methods of the forest industries, but not the use of his home region's forests and swamps for economic development per se. His writings on Southport indicate that he welcomed the development of the little seaside town, which stood to benefit from its proximity to beautiful *and* productive nature. In other words, despite the devastation, all was not lost: evoking an age-old belief in humans' capacity to turn even the most hostile environments to their advantage in successive stages of progressive development, Curtis expected the depleted forestlands of the Lower Cape Fear to become fruitful agricultural fields. He suggested that farmers, in the wake of the forest industry, dispose of their "worthless dogs" and pursue sheep husbandry and truck farming. He envisioned a farming region where small farms, neat cottages, durable fences, and good livestock would herald the disappearance of the razorback hog, the reappearance of trees of all kinds, and the emergence of a profitable and lasting trucking industry that would turn eastern North Carolina into "the garden spot of the world."[16] His prediction echoed the confidence of land dealers who had tried to sell wooded lands on the premise that once trees had yielded their turpentine and their wood, farming the land was the logical next step.[17] It still resonated years later when James Boyd wrote "Old Pines," a story about a railroad through the longleaf pine forest whose proprietor, and his daughter after him, keep finding ways to make forest exploitation pay: when the departure of turpentine and logging crews had brought the trains to a standstill, the prospect of a freshly dredged harbor at the end of the line blew new life into the railroad.[18]

For communities living in close proximity to swamps and forests, the extraction of resources from a changing environment for subsistence or profit was essential to their livelihood and an integral part of their shared lives, and historically such communities forged many, often conflicting relationships with their surroundings.[19] The pursuit of economic gain as well as of opportunities for employment and enterprise; the desire to retain forests, even near built structures, for aesthetic, pragmatic, and recreational ends; the creation and appreciation of certain types of scenic beauty and a sense of loss over their distortion or disappearance; subsistence hunting, fishing, and foraging; the gathering of firewood; the extermination of animals, people, and plants considered useless "vermin" and destructive "pests"; the notion that the environment dictated to some degree the direction of human history; and the belief that nature was God's creation and gift to man—this complex web of thought, belief, and practice shaped, and took shape within, the evolving relationship between human communities and their environment. People who named their churches "Piney Forest" or referred to trustworthy politicians as "Good Timber" gave voice to this historical relationship[20], as did the visitor to the growing town of Lumberton who remarked that "the rugged woods are putting on city airs."[21] Human communities lived in reciprocal fellowship with the surrounding woods, their relationships marked by coexistence, multiple uses, and blurred boundaries between human beings, the built environment, technology, and nature. There existed, in the words of Wendell Berry, no "clear distinction between the living organisms and skills of technology and its mechanisms." Rather, "lives, skills, and tools" being "culturally indivisible" and "all of a piece," communities on the edge of forest and swamp did not experience economic development as being in stark opposition to the natural order of things.[22]

"The only valid way of knowing and learning"

Given the complex character of this relationship, as well as the fact that it evolved long before the rise of environmental science and Deep Ecology, it should not surprise us that the communities in the Lower Cape Fear region long valued "their" forests and swamps from a cultural standpoint more than from an ecological point of view, that for them the abundance of *any* tree species trumped the preservation of particular virgin stands, or that they proved willing to see the ecological downsides of economic development as an acceptable by-product of its material benefits. We cannot fault them for not thinking ecologically, for not perceiving wetlands in terms of their biodiversity and interconnected ecological communities of mutually dependent species, for not appreciating the swamp as wild, unspoiled, and untouched by human endeavor. It would be especially wrong to suggest that because they did not appreciate wild and pristine nature the way today's wilderness advocates do, they did not appreciate nature at all. The relationships swamp and forest communities forged with their surroundings gave rise to a body of traditional, local, practical, experiential, and very much appreciative knowledge about nature that in turn informed a range of local environmental attitudes and practices. To all of this, modern ecology and environmental science are not so much correctives as alternative ways of knowing nature, and as such much more likely to provoke resistance than the tenets of capitalist enterprise, which often meshed far more easily with local worldviews and practices.

The work of Susan Drake Emmerich illustrates the conflict between a locality's traditional environmental knowledge and practices on the one hand, and modern, state-sponsored, ecological expertise on the other. Emmerich, an evangelical environmental consultant, was asked to mediate in a long-standing conflict between the Chesapeake Bay Foundation and

Tangier Island's community of fishermen over the latter's fishing practices in the bay. The watermen, as the islanders call themselves, have historically had a reputation for being lawless, for roughly handling unwanted visitors, and for being suspicious of outsiders in general and of government intrusion into their affairs in particular. During the American Revolution they sided with the British Empire and took to looting American ships, earning a reputation among nearby patriots for being "little more than pirates" who hid in the island's marshes between raids. Methodist preachers who came to the island reportedly had to use their fists to gain an audience and "often preached the gospel with blood running down their faces" until Methodism became so entrenched on the island that "the Tangier Islanders' faith-based worldview governed their outlook and attitudes toward the environment, and the way they viewed relating to neighbors and participating in community affairs."

No wonder that for years secular environmentalists at the Chesapeake Bay Foundation, most of whom "believed that their scientific approach to 'knowing the world' was the only valid 'way of knowing and learning,'" had tried in vain to convince the fishing community on Tangier to abandon its harmful practices of overfishing and dumping trash in the bay. In addition to being outsiders, CBF representatives approached the conflict with an environmentalist ideology that made little sense to people who are "with nature all the time" and therefore "take nature for granted." In the watermen's eyes, drawing a living from the waters of Chesapeake Bay was perfectly reconcilable with their religious beliefs; but the issues of pollution and overfishing, framed as they were in the terminology of outsiders who seemed to threaten the watermen's livelihood, fell outside the scope of their faith. "Due to a near total reliance on the Bay for their livelihood," writes Emmerich, "they have a difficult time with the idea that sometimes the needs of natural creation might take precedence over the needs of human creation."

When Emmerich stepped in as a consultant, she managed to turn the tide by appealing to the community's religious faith and faith-based worldview, rather than by trying to get it to subscribe to the Chesapeake Bay Foundation's ecological discourse. Working from the hypothesis that "faith-based communities require faith-based approaches," she convinced the watermen and their families to enter into a covenant that bound them to environmental practices, reconcilable with their extractive activities, that they came to understand as crucial to their Christian stewardship of God's creation. At the same time, she encouraged "outside professionals . . . to recognize, respect and incorporate the waterman's experiential knowledge and economic and social needs into their decisions and policies."[23]

Janisse Ray's *Ecology of a Cracker Childhood* further illuminates the complex and entrenched character of alternative ways of knowing nature. The *New York Times* hailed the book as the longleaf pine forest's *Silent Spring* because of Ray's careful descriptions of the frail connections between natural and human communities in Georgia's longleaf pine belt. Her discerning sensitivity toward biodiversity and her alarm at its rapid disappearance certainly make for a compelling, passionate account that appeals to modern environmentalists.[24] Yet it is the story of her father that opens our eyes to a different view of the world. Franklin Ray was an apostolic Christian whose reverence for life inspired him to heal wounded animals, forbidding his children to hurt and kill animals and using harsh corporal punishment and austere domestic worship to get his point across.[25] The ecological knowledge Janisse learned in college was alien to him; he was a *bricoleur* who possessed what Ray calls "native genius," the ability "to take the materials and technology at hand and solve complex problems."[26] He ran a junkyard that slowly expanded into a wilderness of steel, rubber, and rust, but he did not conceive of it as pollution or waste—this was handy stuff to have around and use for barter or repairs.

Most profoundly, Franklin Ray credited his struggles with mental illness, his sojourns in the wilderness of the mind, rather than his daughter's academic book-learning, with teaching him:

> 1) greater love for people. 2) greater love for the earth, the trees, the hills, the valley, the streams, the soil, the animals. 3) the future is everything. 4) my wife is me. 5) to love my family. 6) the true value of my sanity, my health, my well-being. 7) to respect our Creator.[27]

This accounting suggests, first, that love of nature does not depend on ecological knowledge but can stem from a deeply personal experience of one's relation to the world. Second, it teaches that attitudes toward nature are woven into a web of human relations (neighborly, familial, marital, parental), religious beliefs, economic practices, and a sense of one's purpose in the world.

The history of human settlements in and around the marshy forests of the Green Swamp region is very similar to that of the Tangier Island community and the towns and farms of eastern Georgia. To be sure, communities around the Green Swamp were home to a more substantial business community of entrepreneurs who saw economic growth and development as desirable and possible, while the Tangier watermen, although integrated to a degree into mainland markets, confined their extraction to the limitations of the Bay and sought more than anything to preserve their existing economy. Also, inhabitants of the southeastern corner of North Carolina took advantage of the longleaf pine's value for naval stores much earlier, much more rapidly, and initially on a much larger scale than those in Ray's Georgia. But settlers around the Green Swamp shared their intimate connection with nature; they faced similar conflicts about what constitutes a "proper" understanding of nature; they had similar worldviews that shaped their attitudes toward nature; similar ethnic and religious backgrounds, as well as similar social relations, shaped their culture; and they faced similar internal and external conflicts over resource extraction and other environmental practices, as well as similar disputes over what constitutes the common good.

"An all-consuming drive for profits"

Historians who write about such rural places and their struggles with communal environmental issues often approach their histories with as rigid a commitment to specific ways of "knowing the world"—economically and ecologically—as CBF officials, anti-globalization groups, or environmentalist activists. The debate among historians of American rural life over the role of capitalism in the formation and transformation of early America, for example, has been marked by a seemingly unbridgeable chasm between, on the one hand, "market economy historians" who claim that the American rural economy was market-oriented from very early on, comprised of enterprising farmers who deliberately produced crop surpluses, and, on the other hand, "moral economy historians" who claim that most farmers were not market-oriented entrepreneurs but family farmers pursuing self-sufficiency. Extensive research on both sides of the divide has yielded a wealth of information about the economic activities of small farmers and petty entrepreneurs around the time of the American Revolution, but it has not produced conclusive evidence as to the nature of the rural economy during the colonial era and the early days of the republic, and interpretations still differ as to exactly what it was that transformed it. Were small farmers pre-capitalists, perhaps even anti-capitalists, whose primary goal was obtaining self-sufficiency, competency, and family security? Or were they accumulative proto-capitalists who ran up against the limitations of the colonial economy? Did capitalism invade and transform a "moral economy" in which relations of exchange did exist but not dominate? Or did it turn semi-isolated market places into an integrated market economy? And what was it that ultimately signaled capitalism's triumph—free wage labor, regional price convergence, deliberate surplus production, a capital market, or the political dominance of a business class?[28]

The first reason these questions are difficult to answer definitively is that historians have indiscriminately applied the "moral economy" concept to a wide range of societies, historical epochs, economic practices, and indigenous cultures, inspired by their unspoken assumption of the concept's transparency and self-evident truth. Indeed, E. P. Thompson went so far as to suggest that the term he "did father . . . upon current academic discourse" had long since lost all ties to the very specific context in which he used it.[29] One of the most influential—and proper—applications of the concept, James C. Scott's *The Moral Economy of the Peasant*, points to the problem inherent in using it in an American context. Scott's concern, as in much of his work, was with the encounter between the indigenous peasant and the foreign or foreign-controlled colonizer, between the traditional, non-western village and the modern nation-state.[30] To ascribe to English and other European settlements in the Americas a similar "moral economy," therefore, is to ignore that the settlers, their animals, their towns, their churches, and their farms were agents of colonization as much as their governments and royal troops. Yet while historians of Native American life have long understood that the European conquest of two continents and the subsequent state of dependency and ultimate genocide of entire cultures is hardy attributable to the practices of a "moral" economy, historians of rural America maintain that the majority of its inhabitants did not share the drive for profit that motivated the powers of Europe to encourage exploration, settlement, and cultivation—and that that drive for profit, in the form of market capitalism, invaded and transformed the American countryside nearly as much as western colonialism ravaged traditional Asian and African cultures.

The second reason that the questions of the "transition debate" are so difficult to answer definitively is that they are predicated on an understanding of capitalism as a finite set of economic conditions that serve to "transform available, but limited, resources into things that we would like to have—economic goods."[31] According to Winifred Barr Rothenberg, a prominent market historian, the capitalist transformation of agriculture was based on a "reallocation of resources to achieve . . . a rate of productivity growth sufficient to sustain [an] exodus of resources . . . in pursuit of higher returns elsewhere."[32] Writing from the other side of the divide, Immanuel Wallerstein similarly argues that "what distinguishes the historical social system we are calling historical capitalism is that in this historical system capital came to be used (invested) . . . with the primary objective or intent of self-expansion."[33] According to these readings, the development of capitalism manifested itself in the commodification of nature and its exchange as capital for reinvestment and expansion, rather than for competency or "mere" accumulation, signaling a shift from "morality" to "market" as the guiding principle of sustained economic activity. As such, the disappearance of the former provokes a sense of loss and nostalgia, for who could possibly object to self-evidently moral economic behavior? The ascendancy of the market, in these terms, can then only prompt lamentations regarding the erosion of both community and nature at the hands of greedy individualists and their equally intrusive and disruptive handmaiden, the modern nation-state. Resistance to the workings of the market on the part of small farmers, artisans, labor unions, women, runaway slaves, and wilderness advocates logically stands out as an affirmation of nature and humanity in the face of dehumanization and environmental destruction.[34] No wonder that "the market" and its infinite expansion of capital divide us into capitalist and anti-capitalist camps—or, by extension, that the nation-state divides us into nationalists and local communalists—with neither side comprehending the other's blindness to what so self-evidently serves the common good.[35]

Yet what Warren Dean characterized as a "capitalist onslaught" on traditional ways of life, human dignity, and the environment has nevertheless had, and continues to find, numerous adherents and proponents.[36] This forces us to investigate how communal understandings of capitalism, of nature, and of community itself have historically enabled people to reconcile the seemingly contradictory values of business, community, and nature within their individual and shared lives; why many of them have taken seriously, when we might not, the claims of companies such as Federal Paper and Council Tool and see them as relevant to, and reflective of, those lives. It is my contention that the market and the nation-state do not rigidly divide Americans into antagonistic camps of capitalists and anti-capitalists, or of champions of national interests and proponents of the local. Rather, they tend to divide people's individual psyches: people have historically wanted the market to shape, partly or entirely, only certain aspects of their lives, while wishing to steer other aspects clear of its influence. Similarly, they have generally wanted the nation-state and its subsidiaries to govern, partly or entirely, only certain aspects of their lives.[37] While rural settlers in colonial America and the early republic may at times have resented and even resisted the intrusion of royal governments and mercantile interests as much as they despised the raids of native tribes or the dangers nature posed, they nevertheless shared with their overlords a desire to tame the American wilderness and transform it into a cultivated landscape of privately owned farms, a desire for acquisition and consumption by means of resource extraction and trade, and a sense that the colonization of the New World was a divinely sanctioned project that served exactly these purposes.

As a result of this embrace of market and state, no matter how tentative or reluctant or incomplete, the local always had a hand in its own transformation by supposedly "outside" forces. As Elinor Melville reminds us, "external developments, such as the expansion of the international market," are not sufficient to explain local economic growth: "local events assumed critical importance in shaping local developments."[38] When people chose to participate in the market to one degree or another, to submit to certain of its demands while trying to avoid others, or to resign themselves to its growing presence, they reconciled within themselves and within their community their place in the market with the aspects of their lives they hoped to keep from being tainted by it. (And the same goes for their response to the power of the nation-state.) This is as true for people who embraced capitalism as it is for those who viewed it with suspicion or even disgust. As Gordon S. Wood writes, moral economy historians "seem to have an overdrawn and caricatured image of an entrepreneur or capitalist as someone who thinks about nothing but the bottom line and has an all-consuming drive for profits that rides roughshod over the needs of his family or his relationship with the community."[39] If capitalism is merely a finite set of economic conditions, no entrepreneur was ever truly a capitalist: they were also, among many other things, fathers, mothers, children, husbands, wives, lovers, siblings, church members, neighbors, citizens, and hunting buddies. Some aspects of their lives in these various capacities may have been market-oriented, even profit-driven; but many were not, yet these constituted crucial parts of their lives and identities. We cannot see their participation in the market and its often detrimental effects on nature and community only in the limited terms of economic practices and the material or capital wealth they generated, and then assume that they, too, had divorced these from the other aspects of their lives.

I do not intend simply to argue hereby that "capitalists are people too" or that "we are all capitalists, we are all moral economists." Rather, I want to posit an interpretation of capitalism, not simply as an economic system or a form of political economy, but as a comprehensive *worldview*: capitalism is a "perceptual framework," a way of seeing the world and human history that extends our common understanding of capitalism into the social, cultural, environmental, intellectual, and spiritual realms of human life that in their totality reconcile what seem to be the contradictory economic and non-economic aspects of life in modern capitalist societies.[40] Its fundamental premises are that households, communities, societies, and really the world as a whole can and should be organized in such a way that economic growth and development are potentially limitless; that from economic growth and development flow increased democracy, wealth, and freedom to an ever-growing number of individuals; and that this widespread, continuous, and accretive improvement of the quality of human lives, if left to proceed without too much interference, constitutes all but inevitable progress. It finds its ultimate expression in Milton Friedman's influential writings on individual freedom, in the popular "free market theology" of mega-church evangelicals, in Abraham Kuyper's defense of Calvinism as a "life-system," and in the objectivist cult of Ayn Rand that poses capitalism as the *only* moral economy in history.[41]

As a worldview, then, capitalism not only incorporates the ideals of classical liberalism and neo-classical economics, it also subsumes the humanistic ideal of the autonomous, individual self freely pursuing his or her interests, restrained only by the interests of others; modern religious beliefs that conflate the purpose of the self, the family, the church, and the nation; the liberal political ideology of natural rights; and the materialism of the natural sciences—all are subject to the logic of the market and indeed well served by its drive toward efficiency and profitability. Free wage labor, private property, capital reinvestment, technological innovation, and limited government interference are therefore not the defining characteristics of progress but merely the necessary mechanisms by which to achieve it. They pose no threat to family, community, church, leisure, art, civic life, or nature; instead, as long as these are organized according to the capitalist worldview's self-evident premises of what is good and desirable, these mechanisms serve their greater good. The capitalist worldview, after all, envisions not a world of destruction and exploitation but a prosperous and satisfying way of life that includes those benefits of family, nature, and community that reflect its premises. It is not anti-family, or anti-nature, or anti-community, as long as these do not pose an obstacle to the freedom of the autonomous individual to pursue his or her happiness in an unfettered market economy.

At the same time, the promise of inevitable and unlimited progress justifies as "only temporary" the hardships, non-freedom, injustice, and heavy-handed government intervention people and nature may suffer. And its denial of possible natural constraints to progress not only gives it somewhat of an extra-physical façade that reinforces the idea of "the market" as a free-floating, autonomous force but also masks the physical degradation that is the downside of its success.[42] Only an understanding of the workings of this political economy as the attempted realization of a worldview that not only sanctifies rapid economic expansion but also cloaks its destructiveness in an ideology of wholesale social, cultural, and spiritual progress can properly explain its longevity, widespread pursuit, and continued appeal. Such an understanding can explain why communities around the Green Swamp, rather than tremble at the prospect of the social fabric's unraveling when the *Southport Leader* called for "the magic wand of capital and enterprise" to unlock the potential of the area's natural riches, eagerly anticipated an "awakening along all lines of industry" to enrich and improve their rural life.[43]

"The military-industrial conspiracy to murder Creation"

All the while, a community's understanding of itself as bound by common interests and a shared identity, of its economic practices and their benefits for the community, and of its place in nature can mitigate the detrimental and damaging effects of unbridled economic development. As Martin Bruegel shows in *Farm, Shop, Landing*, the "ordinary people" of the rural mid-Hudson Valley who "participated, eagerly or reluctantly, in the construction of a market society," in large part through the abolition of manorial estates, did neither wholeheartedly embrace or entirely reject the premises and workings of the market economy. "Zest and frolic," as well as religion and social rebellion, were not alternatives to the market, but the means by which rural New Yorkers negotiated the tension between liberty and equality that characterized the development of a capitalist market society.[44] In order properly to understand the history of socioeconomic development in an area like the Green Swamp region, therefore, we must heed Christine Rosen and Christopher Sellers' call for an "eco-cultural" approach to business history. Only such an approach, sensitive to the ecological (nature), the cultural (community), and the economic (business) aspects of both profit- and subsistence-driven landscape transformation, can properly embed business enterprise in the complex web of cultural, social, and environmental practices in which it is always entangled and restore nature, from its traditional passive role as the inanimate setting for vibrant capitalist development, to its active part in that web as well.[45]

The writings of Wendell Berry are among the most profound evocations of community-centered rural life as both a victim of, and an antidote to, the ravages of modern capitalism. Exemplary of practical ecological thinking within a rural context, his case for local economies, family farming, and individual and communal responsibility for respectful stewardship of the land, as well as his denunciation of industrial farming, sprawling cities, large-scale business, and selfish individualism, are essential reading for anyone who fears that suburbs, strip malls, and superstores will soon bulldoze their way over every last bit of local culture and wild nature. His work as an essayist, poet, and farmer is especially poignant because it is rooted in a comprehensive worldview that, as a disconcertingly obvious alternative to the dominant capitalist worldview, is now as radical as it is rooted in tradition. As a Christian who believes that the earth is God's creation and the people who inhabit it its stewards, Berry argues that what people are for is not to live in mega-cities so they can work as faceless cogs in the military-industrial machinery, alienated from the land that sustains them. Rather, it is to live on and with the land in healthy, local communities that nurture all life. Berry denounces the suffering inflicted on land and people, not in ecological terms that separate humans from nature, but as "the most horrid blasphemy" against a singular divine creation of which humans are an integral part. His is a Christian worldview that sees family farming, communal life, and the church's witnessing as proper ways to honor God and His creation, as institutions that cannot and should not to be shackled to the capitalist machinery of "progress."[46]

Yet Berry's work also points to the problems of seeing "community" as a panacea for the ills of modern society. While an eco-cultural approach to business history will alert us to the mitigating effects healthy communities have had, and continue to have, on "capitalism's conquests," we must be careful neither to romanticize the rural community nor to take its contours and cohesion as self-evident, fixed, and organic.[47] Nor should we dismiss as disingenuous capitalists' alternative visions of community: "Communities are to be distinguished," wrote Benedict Anderson, "not by their falsity/genuineness, but by the styles in which they are imagined."[48] Berry made as much clear when he addressed the students and faculty of the Southern Baptist Theological Seminary in Louisville, Kentucky, an audience we might expect to play choir to Berry's preacher. Berry expected otherwise, and proceeded to demonstrate that environmentalists' indictment of "the culpability of Christianity in the destruction of the natural world and the uselessness of Christianity in efforts to correct that destruction," while perhaps cliché, was in many respect just:

> Christian organizations . . . remain largely indifferent to the rape and plunder of the world and of its traditional cultures . . . Most Christian organizations are as happily indifferent to the ecological, cultural, and religious implications of industrial economics as are most industrial organizations . . . [and] just as likely as anyone else to join the military-industrial conspiracy to murder Creation.

Berry also criticized the "conservationist indictment of Christianity" as indicative of an inadequate understanding of the Bible and "the cultural traditions that descend from [it]," but what is important here is this: the church, the *community* of believers that ought to carry its cross and witness to the evils of rampant capitalist accumulation and exploitation, that imagines itself willing to die for the Gospel, instead "folded its hands while so much of the work of God was and is being destroyed."[49] The church community did not properly imagine itself as a transnational body of disciples opposed to the murderous combination of market forces and nation-states.

Berry's harangue alerts us to three aspects of community that undermine a generally uncritical application of the concept. First, people are always part of several imagined communities at once, which means that they construct for themselves, and others construct for them, different, often overlapping, and sometimes conflicting identities as community members. It also means that several imagined communities make demands on their loyalty: people may imagine themselves Christians, but they are also Americans, Southerners, and North Carolinians who furthermore belong to such communities as Brunswick County, Southport, their family, their race, their gender, and so forth. To Berry it is painfully obvious that American Christians have generally privileged their identity as citizens of a powerful, capitalist nation-state over their discipleship.

Otherwise, they have come to understand their discipleship as part of, if not subsumed under, their national identity, which points to a second problem with the concept of community: their identities can be folded—imagined—as secondary, tertiary, or in any case auxiliary identities into the primary communities to which people pledge allegiance. The idea that the United States is a nation of special concern to God is a manifestation of this, as is the idea that the material comfort its citizens enjoy constitutes the blessings bestowed on "God's Country," which therefore should not stand in the way of further material improvement; the notion of "the market" as either an autonomous realm or God's instrument of global liberation that impinges in no way on one's individual relationship with Christ is another. As such, Christianity has been thoroughly co-opted into the capitalist worldview, with much of the church—or, equally problematic, many churches—now serving "progress" rather than the kingdom of God while deluding itself with the belief that they are one and the same.

Moreover, the ubiquity of this worldview among the branches of the Christian community points to a third problem with the concept of community: communities are held together by power structures and by political relations that dictate the terms of belonging to the imagined community, define its collective identity, and police its boundaries.[50] What Berry implicitly criticized in his lecture—and explicitly elsewhere—is the complacency of too many Christians not only in the face of the military-industrial conspiracy's expansion but also in the face of church leaders who deny them the full Gospel and instead affirm them in their capitalist and patriotic ways of life.[51]

The recent collection *Forest Communities, Community Forests* clearly illustrates the pitfalls of taking "community" at face value. The premise of the case studies it contains, one that is hard to dispute, "is that healthy ecosystems depend on healthy communities and vice versa."[52] Their authors seek to show how communal participation in the rebuilding and management of forests benefits the health of both nature and community at a time when "efforts in the United States to reduce public participation in decisions about public forests returns us to the practices of favoring a few at the expense of communities and the environment."[53] Communities that live in close proximity to forests, they argue, do not conceive of the environment as a separate entity on which human beings can and must from the outside impose this will or that; rather, they see the forest as part of their community, and their community as part of the forest. One study depicts the attempts of Catron County, New Mexico's government to "take control of Forest Service lands within its jurisdiction in order to manage them for commodity utilization based on the principle of 'custom and culture'"; the county's intent was to address the impact of federal decisions on "traditional ways of life or historic, social, and economic community practices."[54] Another study shows how "corporate and administrative boundaries that overlie the physical landscape" of the Upper Swan Valley in Montana "make land management and decision making not only complex but also potentially explosive between neighbors, absentee owners, and others whose decisions affect the land."[55] A third study shows the success of certified sustainable forest management in Aitkin County, Minnesota, where "a local entrepreneur, county-level land managers, and local and nonlocal nonprofit organizations . . . were able to meld their ideas about good business, community development, and environmental responsibility with community and north-woods restoration and sustainable management of the local forest."[56]

In addition to their attention to successful attempts to reshape forest management and community building on a local level, however, these studies share a problematic characteristic: they all take the existence of "community," largely on the county level, as glaringly self-evident, paying little attention to the possibility that "the practices of favoring a few" may very well happen *within* a given community. While their authors depict federal boundaries and activities as intrusive, exclusive, and potentially harmful, they take county boundaries as proper delineators of "community," even as they run through forests, larger ecosystems, and alternative configurations of community. Moreover, the ultimate purpose of the collection is nevertheless to reflect the official objective of the Seventh American Forest Congress: "to develop a shared vision, a set of principles and recommendations that will ultimately result in policies for *our nation's forests* that reflect *the American people's vision* and are ecologically sound, economically viable, and socially responsible."[57] Laudable as this goal may be, it glosses over the inherent conflict between the possibility of a singular vision shared by "the American people" on the one hand and the needs and desires of local ecological and human communities on the other. A vision shared by *some* American people in one locale and imposed on others through a particular system of government as a defense of "American interests" may very well come at the expense of *other* American people and their visions, not only in other locales but within that same locale as well. The power to determine the identity or the policies of any community, after all, has rarely been distributed equally.[58]

"A detailed 'map' of its terrain and its people"

In this eco-cultural business history of the Green Swamp and its environs, then, I argue that while communal values and practices at times mitigated the environmental and human ravages of capitalist economic development, communities' cultural identities were nevertheless imagined, constructed, and maintained in ways that facilitated capitalist enterprise and directed the use and transformation of nature and community in accordance with the ideological premises of the capitalist worldview. For this purpose I rely greatly on James C. Scott's concept of legibility as "a central problem of statecraft":

> The premodern state . . . knew precious little about its subjects, their wealth, their landholdings and yields, their location, their very identity. It lacked . . . a detailed "map" of its terrain and its people . . . a measure . . . that would allow it to "translate" what it knew into a common standard necessary for a synoptic view.

In response, the modern state devised ways of seeing and knowing that were abstract, simplifying, scientific, and standardizing; that translated into abridged, "cadastral maps" that allowed high-modernist governments and their cadre of experts to see what was now a legible landscape exclusively in terms of their own narrow interests and subsequently impose that vision on the land and its people. The forceful creation of scientifically managed forests, fully planned capital cities, collective industrial farms, and new villages, however, came at the expense of natural diversity and resilience, as well as of "practical knowledge, informal processes, and improvisation in the face of unpredictability" on the local level. This complex jumble of knowledge and practices stood in the way of the grand, holistic, fully planned schemes state visionaries preferred, so they, to at best mixed and never complete success, sought to brush it aside as "backward," "unscientific," and "illegible."

While in and around the Green Swamp desired degrees of legibility were not imposed on traditional peasant societies by colonial rulers, by post-colonial, westernized governments, or by heavy-handed socialist states, the concept of legibility and the history of its implementation, Scott warns, are "as applicable to market-driven standardization as they are to bureaucratic homogeneity." While settlers around the Green Swamp did not have the kind of practical knowledge of local nature that peasant communities in Africa and Indonesia had passed on for thousands of years (and that the high-modernist state ignored or hoped to obliterate), the legibility that in *Seeing Like a State* came from without and above in southeastern North Carolina also came from within as settlers and their governments all sought to colonize an unknown, illegible landscape but disagreed about the proper ways to do so. In this situation, the kind of legibility favored by governments was rarely diametrically opposed to the kind of order and organization favored by a community. Rather, preferred formats and levels of legibility were communally defined and contested: many saw agents of colonization such as the state, the plantation, business, the church, and the city as the proper manifestations of the order their worldview prescribed and accordingly aligned themselves with those institutions; others resisted, partly or wholly, their dominant position or their methods of imposing legibility.

The starting point for *Seeing Like a State* (which Scott ultimately abandoned) was the troubled relationship between the modern state on the one hand and people without a fixed, permanent residence—gypsies, vagrants, the homeless—on the other. Long before the modern state became today's behemoth, however, with all the tools at its disposal to track such people down and bring them in line, communities around the Green Swamp already shared with distant and nearby governments a strong suspicion of the wild, the rootless, and the lawless—be they people, animals, or the forces of nature. And they extended this suspicion not only to the world around them, but also to an emerging capitalist market whose wild fluctuations, rogue "players," uncertainties, and seemingly autonomous forces made it as much an illegible wilderness as the Green Swamp itself. In some cases, communities and individuals sought local solutions to what they considered pests, vermin, and other untrustworthy creatures; in other cases, they petitioned the state for help; in other cases yet, they sought what we might call "market solutions"—entrepreneurial attempts, really, to wave the magic wand of capital and make the wild, in all its manifestations, legible and navigable. In all cases, their responses to nature, to recalcitrant people, to alternative ways of seeing the world, and to the market economy, prescribed and affirmed community values regarding who was in and who was out, who could be trusted and who could not, which practices were proper and which were improper, which knowledge was right and which was wrong.[59] So as much as they are still natural wonders, the forests and wetlands of the Green Swamp region, mosaics of legible and illegible nature, thus reveal the conflicted and contested responses of human communities to the challenges of nature and business.

1

"A Country But Wild and Imperfect"

One still easily spots the Green Swamp on recent road maps of North Carolina: located just north of the town of Supply in Brunswick County, the state's southeastern-most county, the place looks empty, devoid of human alteration. There, along North Carolina state highway 211, the Nature Conservancy owns a preserve of more than 15,000 acres of rare swampland and pine forest, much of which is thought to be in a relatively pristine state. It is one of the few places in North Carolina, if not the entire United States, that still hosts stretches of virgin longleaf pine savanna.[1] To the northwest of the Green Swamp, in Columbus County, lies Lake Waccamaw, another rare natural environment, whose southern shore is part of Lake Waccamaw State Park. With the exception of NC 211, which cuts through the Green Swamp, the signs of human habitation lie on the outer edge of this swath of swamp, lake, and forest. Towns are situated along roads that form a human-made perimeter around the wilderness, but the swamp itself appears largely untouched. Outside this perimeter, in the remaining parts of Brunswick and Columbus as well as in neighboring Bladen, Robeson, and New Hanover counties, the landscape is still rich in protected areas, scenic rivers, and swamps; but it shows a much less distinct separation between human-made and natural environments. This surrounding area constitutes an outer belt around the Green Swamp and its immediate environs, one that brings into sharp relief the swamp's apparent emptiness.[2]

Depicting the Green Swamp in such a way as to emphasize its desolate character, a modern road atlas reflects a specific, narrow way of seeing the land and represents a deliberate simplification of the landscape's complexity for the specific purpose of legibility: it helps one find one's way by automobile among the features of the human-built environment. Upon closer inspection of the landscape that the map describes, the lines between the built environment and the natural environment turn out to be far less distinct. While homes, farms, and logging roads that lie along

Figure 2. The edge of the Green Swamp off NC 211. (Photograph by author, 2006.)

NC 211 may not register on this type of road map, they are among the many human-made structures that reach deep into the swamp. Marked trails cut through forest and swamp to lead hikers to docks, campsites, and shelters; signs posted on the outskirts of thick timber stands warn people to stay out. These are part of the "human footprint" that reveals most landscapes to be neither entirely human-made nor all natural. This is a far more complex mosaic, a patchwork environment in which the human-made blends with the natural to a great many degrees: sometimes the former appears to obliterate the latter, sometimes the latter appears to dominate; sometimes they appear to be in harmony, sometimes in tenuous coexistence; sometimes they clearly clash. Because human-made environments, like humans, are made of the stuff of nature and connected to it by a complex web of human-made and natural relationships, nature is present in even the most artificial of environments. The wildest environments, in turn, are defined and delineated by the very human-made structures that circumscribe them—what appears natural may very well be the product of human intervention, such as the wild plants that thrive in roadside ditches and fireplow lines.

"A sloshy union"

In terms of its ecology, the Green Swamp and its environs are situated in the heart of what used to be the longleaf pine belt of the Atlantic Coastal Plain: the stretch of coast that runs from northeastern Florida to southern New Jersey, some 100 miles wide and once covered with seemingly endless stands of longleaf pine. South of the Delaware River, the plain, whose elevations rarely reach 1,000 feet, consists of a series of broad, step-like terraces and connecting escarpments that slope gradually toward the ocean. On the plain's western edge, the Fall Line, once the eastern continental coastline, divides coastal plain from Piedmont, where sand hills predominantly support a growth of pine forest and scrub oak. To the east lies the maritime environment of the Atlantic Ocean. The subtropical climate has long ensured high levels of precipitation and heat, and this wet and warm climate helped the formation, over millions of years, of an organic mat covering and a thin organic/mineral surface soil from which iron and aluminum leach into the subsoil. Most common are sandy loam and silty loam, which are slightly acidic and low in both organic matter and soluble nutrients. The coniferous and mixed forest growth that these soils came to support further aided the formation and distribution of this soil type. Most notable was the profusion of the longleaf pine, which, with its deep taproots, dominated the vast stretches of flat sandy soil as long as frequent fires kept other growth in check while allowing the fire-resistant longleaf seedlings to sprout. Fire also shaped the undergrowth of turkey oak, scrub oak, and wiregrass that benefited from frequent burns and from the sunlight that reached large parts of the forest through the canopies of spacious longleaf tree stands.[3] "This plain is mostly a forest of the great long-leaved pine, the earth covered with an infinite variety of herbaceous plants," wrote William Bartram in 1791 as he beheld "extensive savannas, always green, sparkling with ponds of water" on his journey through the Lower Cape Fear region.[4]

The slow drainage that is a product of the plain's gentle slope and the inward-pushing ocean tides gave the abundance of water on the coastal plain a crucial role in the shaping of its ecology. In North Carolina's Inner Coastal Plain Region, southwest of the Cape Fear River and southeast of the Fall Line, several main waterways—fed by many swamps, lakes, bogs, and smaller tributaries—made their long, winding way to the Atlantic to create this magnificent landscape. The Cape Fear River runs in a southeasterly direction through Bladen County and turns southward to divide Brunswick County from New Hanover County. This brownwater river is tidal for its entire length along Brunswick County and into Bladen County, and the water is brackish to saline close to the ocean. The Lumber River, a blackwater river like the rest of the region's rivers, runs southeast through Robeson and then turns southwest toward South Carolina, dividing Columbus and Robeson counties before meandering to Winyah Bay in South Carolina.[5] The Waccamaw River originates in Lake Waccamaw in Columbus County and slowly winds its way south to Horry County, South Carolina. Its floodplain contains several landform features not found elsewhere in the county; geologists claim that the Waccamaw originated as a brownwater river in the Piedmont, but about 75,000 years ago an uplift along the Cape Fear Fault "beheaded" the river and created Lake Waccamaw as its new source. This would explain the river's small size as well as the large expanse of its floodplain and the landform features more common to brownwater floodplains.[6]

This ocean-bound, fluvial web cut through the longleaf pine belt with its widely dispersed longleaf pines and dense undergrowth of wiregrass and Carolina dropseed, creating a "sloshy union" of extensive longleaf pine savannas, dense tidal marshes, pocosins, peat bogs, and Carolina bays.[7] In the ecological interstices—or ecotones—between these systems, the longleaf pine forest gave way to mixed pine stands that included loblolly, shortleaf, and pond pine. Across the floodplains, these mixed pine stands stood side by side with a wide variety of shrubs and grasses as well as the cypress, cedar, juniper, white cedar, and swamp tupelo that stood in and around the many swamps, bogs, and bays.[8] Along the ocean shore, streams that emptied into the ocean-fed estuaries and coastal wetlands while the ocean's tidal flow fed saltwater inlets and tidelands. In southeastern Brunswick County, the Shallotte River and Inlet and the Lockwoods Folly River and Inlet constituted a transitional zone of tidal marshes with oak, black needlerush, and salt grass between the freshwater coastal plain and the ocean. Large tidal marshes also formed in the tidal area behind the barrier islands off the Atlantic coast, especially where the rivers emptied into the ocean. An inter-dune pond community of common reed, black needlerush, cattail, bulrush, and spike rush; an oceanside dune community that hosted unusual amounts of bitter panic grass; and communities of cabbage palm, wild olive, Carolina laurel-cherry, and red bay—all formed additional rare ecosystems.[9]

The rich mosaic of habitats hosted an abundance of wildlife, ranging from large predatory mammals, reptiles, and birds that thrived in the expanse of habitat, to medium-sized and smaller animals that found plenty of sustenance there, and to little creatures that carved out a niche in this ecosystem of water, woods, and fire. Black bears, red wolves, panthers, and alligators prowled the Green Swamp region in large numbers, as did deer, otters, raccoons, bobcats, squirrels, and rabbits. There were venomous and non-venomous snakes that never failed to capture the frightful imagination of visitors to the swamp, turtles, and many frog species whose nocturnal cacophony still resonates across the swamps. Eagles, owls, and peregrines soared above the swamps and forests in search of prey, while turkeys trod through the undergrowth, cranes kept close watch over streams and inlets, turkey vultures swooped to carcasses left in the woods, and bobolinks and other bird species infuriated or delighted human settlers. A great variety of fish species could be found in the abundance of water, which included the Atlantic; situated on the environment's eastern edge, the ocean was, in many ways, an environment unto itself, but it supported species as diverse as sharks, loggerhead sea turtles, oysters, eels, and shrimp that in their existence close to the coast were important components of the ecological exchange between shore and ocean.[10]

In addition to the proliferation of large, often edible species that early European settlers to the area were quick to notice and pursue, the interaction between water, soil, vegetation, climate, and fire, which over the course of millions of years shaped this section of the coastal plain, produced a significant number of rare natural communities in which rare and less visible species found a perfect habitat. Flatwoods salamanders and gopher frogs thrived in the little ponds and potholes amid the pinewoods bluestem and toothache grass of the wet plain, while red-cockaded woodpeckers preferred to excavate nesting holes in old longleaf pines sufficiently infected with red-heart to soften the heartwood. Other species sought shelter high in the trees or in tortoise and snake holes, drawing their shelter and sustenance from fire-dependent tree, plant, and grass species.[11] These communities of mutually dependent plant and animals species had emerged over time as they adapted to nutrient-poor soils, frequent burning, wet and warm summers, and varying levels of precipitation, inundation, and saline water encroachment.[12] Because the warm waters of the Gulf Stream bend away from the coast north of the Cape Fear, southeastern North Carolina is the most northerly habitat for many species that do not tolerate colder weather. As a result, the area supports the most diverse flora and fauna as well as some of the highest densities of natural diversity north of Florida. Richard LeBlond of the North Carolina Natural Heritage Program has identified thirty-six different natural community types and subtypes in Brunswick County alone, calling the county "a center of species diversity along the Atlantic Coast," while Columbus County is home to twenty-six such types and subtypes.[13]

The most notable feature of this landscape was the Green Swamp, a large elevated or "domed" peat swamp that today covers one-fifth of Brunswick County's land surface and extends northward into Columbus County. Brunswick's land surface consists largely of unconsolidated sand, silt, clay, and peat in various combinations; extensive pocosin areas such as are found in the Green Swamp region are among the highest elevations and most prominent physiographic features. As a result, water flows in all directions from the elevated peat land, and the swamp is headwaters for streams flowing northward and eastward into the Cape Fear River, westward into the Waccamaw River, and southward into Lockwoods Folly and Shallotte rivers. Because of the abundant yet dispersed water runoff, the swamp supports large swaths of open longleaf pine savanna, punctuated by dense, sometimes impenetrable stretches of pine flatwoods, pond pine woodland, pocosin, and bay forest areas that all host an amazing diversity of flora and fauna.[14] Scientists working in the Green Swamp have found "as many as forty-two vascular plant species in ¼ square meter plots, as many as fifty-two in one square meter, and up to eighty-four species in plots of 625 square meters," among them pitcher plants and other rare insectivorous plants.[15] In 1995 when biologists considered the large remnant of the original swamp still to be in more or less natural condition, the swamp supported populations of twenty-six rare plants and sixteen rare animals, including the eastern cougar, Bachman's sparrow, the Cape Fear threetooth mollusk, the Venus flytrap, and Carolina grass-of-parnassus. Some of the smaller species, such as the Greenfield rams-horn snail and the Magnificent rams-horn snail, are known to exist only in this corner of the world. The swamp also hosts many of the larger animal species that are typically associated with this type of wilderness, such as black bears, water moccasins, and alligators.

Another striking feature of the landscape that surrounded the Green Swamp were the Carolina Bays, elliptical depressions usually oriented along a northwest/southeast axis, and surrounded by a low sand ridge called a bay rim. The size of the Carolina bays, of which there are over 20,000 on the Atlantic Coastal plain, ranges from a few to thousands of acres.[16] They usually depend completely on rainfall for their water supply, and their name derives from the abundance of sweet bay, loblolly bay, and swamp red bay trees around their edges.[17] Stands of pond pocosin, pine, and Atlantic white cedar also surround bays that have not been drained, which suggests that prior to widespread drainage the vegetation in bay areas could be as dense as that of the Green Swamp. Bladen County is especially rich in Carolina bays, the largest being Singletary Lake, Jones Lake, Salters Lake, and White Lake. Around these bays, fetterbush, pepperbush, gallberry, huckleberry, and sheep laurel also thrive in the acidic soil, as does the rare white wicky, a relative of mountain laurel. Carolina anoles, fence lizards, spring peepers, southern leopard frogs, and cricket frogs are abundant, but one can also find the rare Brimley's chorus frog and the endangered pine barrens tree frog among the reptilian life. Carolina wrens, chickadees, and black vultures are still common; the yellow-throated warbler, the white-eyed vireo, the pileated woodpecker, the red-tailed hawk and the endangered red-cockaded woodpecker are much rarer now.

The largest of all water-filled bays in North Carolina is Lake Waccamaw. The lake and lakeshore constitute as rare an environment as the Green Swamp immediately to its southeast. The lake and the swamp share many natural community types, and they strike the casual observer traveling from the one to the other as one sloshy environment. The lake and its immediate environs are at the same time the exclusive habitat of several fish, mussel, and clam species.[18] Near the lakeshore, the water supports one of the most diverse and best developed natural lake shoreline plant communities in the state. Several populations of the Cooley's meadowrue, a flimsy herb that can reach two meters, grow in savannas and in bog communities that also include tulip poplar, cypress, and Atlantic white cedar. It sprouts a ball-shaped, pinkish flower and depends on some kind of disturbance, such as fire, for survival. Natural scientists have found only twelve populations of this herb, in southeastern North Carolina and Florida. Along the Waccamaw River, which springs from the lake at its southwestern edge, lies the Upper Waccamaw River Swamp, where black bears, snakes, and alligators add to the wild beauty of this rare environment.

Figure 3. A mixture of dense growth and water on the southern shore of Lake Waccamaw. (Photograph by author, 2006.)

"Dey do say de hants walk dat place"

Is it any wonder, given our growing knowledge and understanding of their incredible biodiversity and important hydrological function, that wetlands such as the Green Swamp have in recent decades generated such passionate and widespread mobilization of environmentalist activists? Or that SWAT ominously warned of a "Wetland Genocide in the South," joining forces in the late 1990s with FOGS in a tenacious fight to prevent the proposed building by Riegel Ridge Corporation of a large landfill in the Green Swamp?[19] How could anyone ignore the obvious and turn a blind eye to the fact that draining and filling in these wetlands for the purposes of building landfills and golf communities devastate very fragile ecosystems, rich in rare and endangered flora and fauna and crucial to natural patterns of water flow and runoff?

It is common to blame autonomous and anonymous market forces for such destructive environmental change, and certainly not unfounded to point to the avarice of individuals and corporations that have ridden roughshod over human and natural obstacles to the fulfillment of their material desires. Yet such explanations tend to reflect the inability or unwillingness on the part of environmentalists to see their knowledge of nature, as well as the ideology and activism it engenders, as anything other than logical and self-evident truths that competing interests simply choose to ignore. Widespread awareness of the complexity of fragile ecosystems, an appreciation of their wild beauty, and the idea that human beings are at best highly developed organisms *within* them—equal to all other life forms—or at worst a destructive blight *on* them are relatively recent outgrowths of modern environmentalism. Yet their proponents have been quick to dismiss as irrelevant, misguided, and harmful the worldviews, beliefs, and experiential knowledge that people have much longer brought to bear on the world they inhabit and exploit. In their efforts to impose on human interactions with nature *their* vision of "sound" environmental practices, they have sought, as much as the capitalists they deplore, to dictate the "correct" terms of debate, claim a monopoly on the "right" way of knowing nature, and appropriate the political power to force the implementation of "proper" practices on nature and humans. They have also been as quick to depict their opponents as greedy, exploitative capitalists as those who espouse capitalism have simplistically demonized environmentalists as "tree huggers" or even "eco-terrorists."[20] They have failed, though, to displace traditional notions of wetlands as unproductive wastelands, as treacherous places of danger and disease, as illegible and undesirable landscapes. More important, they have failed to breach the equally entrenched view that nature and community, rather than victims of selfish, profit-driven business, are reconcilable with, if not well-served by, the hegemony of the market.

"Between the sixteenth and seventeenth centuries the image of an organic cosmos with a living female earth at its center gave way to a mechanistic world view in which nature was reconstructed as dead and passive, to be dominated and controlled by humans," Carolyn Merchant argued in *The Death of Nature*.[21] But this paradigm shift, crucial though it was to the eventual development of the economic theories of Adam Smith and Karl Marx, appears to have occurred mainly in elite academic circles. For most human beings, the workings of the world around them long continued to warrant fear, wonder, dread, and awe; they were subject, not to rational explanation, but to mystified, imaginative, myth-based storytelling about a universe not of man's own making.[22] From the earliest days of European settlement, for example, North America's "dismals" and their wild inhabitants struck fear in the hearts of settlers accustomed to organized, productive landscapes. Swamps "violated their norms of orderliness and presented an incomprehensible, chaotic landscape." These were "evil," "wicked," and "sinful" places, dwelling places of Satan and his Native American minions. From a religious viewpoint, they were good for trials of faith and errands into the wilderness, for the cleansing of individual souls and the apostate church. Permanent habitation, however, required far-reaching transformation, yet motivated by the same understanding of the world. Settlers vilified swamps in sermons, correspondence, oral legends and myths, and formal literature; they avoided them when they could, and when they could not, they sought to make them hospitable and productive.[23]

The wildness and seeming impenetrability of the Green Swamp region gave rise to this kind of narrative—to legends, stories, and myths that reflect the region's unique ecology not nearly as much as they do the enduring influence of that original sense of fear, loathing, and wonder. Despite our growing awareness of the area's ecological fragility, these stories continue to have currency, to capture the imagination, and to inspire efforts at cultivation. In part, they reflect the persistence of fearful fascination with wild places and wild things, especially when the wild ventures outside the bounds of a distinctive "wilderness experience" and encroaches on our daily lives. In part, these tales justify further taming the wild with residential and business developments, even as proximity to nature increasingly serves as a selling point for real estate. And in part, they serve as a means to stimulate outdoors tourism and attract a movie industry ever in search of spooky, ghostly, and threatening scenery. As such, they continue to complicate public responses to issues environmentalists wish to cast in strictly ecological terms. They counter scientific and supposedly rational arguments with deeply felt experiential and mythological traditions that are deeply rooted in a fundamental desire to transform the useless and frightful wilderness into useful landscapes, to make no man's land into good man's land.[24]

While popular myths regarding wondrous landscapes may never have struck European settlers in and around the Green Swamp entirely as objective truth, they nevertheless reinforced traditional ideas about the interconnectedness of wild nature and their native human inhabitants. According to a mythological story popularly known as "The Legend of the Lake," Lake Waccamaw was once a garden full of all the world's flowers, guarded by an Indian princess named Waccamaw. Every year, representatives of Native American tribes would come to the garden to receive from the princess a blessing and a wild rose so that their people might prosper. The chief of one tribe, Ashbow, wanted the beautiful Waccamaw to marry his son, but she refused because of her promise to stay in the garden forever. The scorned prince, vowing revenge, waged war on Waccamaw's people and demanded the princess' hand. At the hands of the angered Great Spirit, however, the princess' enemies perished. The garden then sank into a lake and became Lake Waccamaw—to this day, and for this reason, the site of bounteous exotic flora.[25]

The Carolina bays also fire the imagination of observers who, dissatisfied with a scientific explanation of their origin, seek in the realm of outer space a more compelling reason for their existence. The mystically inclined, sometimes backed by outdated or unidentified scientific research, continue to entertain the possibility that the bays formed as a result of extraterrestrial forces, speculating that meteors, comets, or asteroids remain "a viable alternative worthy of further consideration."[26] *The Film Junkie's Guide to North Carolina*, for example, alerts movie buffs who wish to visit the state's famous movie sites to "the largest of the *mysterious* Carolina bays—oval depressions that some scientists believe resulted from meteor showers."[27] The North Carolina Division of Parks and Recreation enjoins visitors to "Venture to Jones Lake State Park and view one of the greatest geological *mysteries* of the eastern United States—the phenomenon of the Carolina bays."[28] To be sure, few geologists pay serious attention to the meteor shower theory, but keeping the possibility of the bays' extraterrestrial origins alive in the public imagination continues to enhance the mysterious quality of the coastal landscape.

Another legend, "The Riddle of Shallotte Inlet," plays on the supernatural as well as on the interconnectedness of Native Americans, wildness, and nature. As folklorist Charles Harry Whedbee explains, Shallotte Inlet on the coast of Brunswick County "is a most peculiar phenomenon." The inlet contains "the Indians' legendary healing waters," which have miraculous powers: "It can, and it does, heal old wounds that have festered and 'wept' for long periods of time." Should rational readers be tempted to dismiss this claim as so much hoodoo-voodoo, the author assures them that "old-time bear hunters have known about the spa for years." This is the kind of phenomenon, after all, that transcends scientific understanding; knowledge of such matters is the traditional preserve of Indians and backwoods hunters—people who knew nature practically and experientially, who were in touch with the medicinal folklore of an area that had seen, as had many coastal settlements, its share of incurable and inexplicable diseases.[29]

One of North Carolina's most enduring tales of mystery and imagination is "The Maco Light," which continues to find its way into travel accounts and collections of North Carolina folklore and ghost stories. The story recounts the fate of train conductor Joe Baldwin, who in 1867 was riding in the rear of a train heading west from Wilmington when his car came uncoupled at the small station of Farmer's Turnout on the Green Swamp's northern shore. He walked out onto the tracks, waving a lantern to warn an oncoming train of the car standing still on the tracks. Ignoring the light, the oncoming train crashed into Joe, decapitated him, and sent the lantern flying into the swamp. Ever since, people claim to have seen the appearance of a flickering light along the railroad track and in the swamp close to where the accident occurred; and legend has it that it is the ghost of Joe Baldwin with his lantern, looking for his head.[30]

The Green Swamp has also been the setting for historical narratives that cement the swamp's reputation as a place of refuge for outlaws, runaways, and renegades while reinforcing the idea that people who successfully take shelter in such a forlorn wilderness must themselves be wild and savage. When Minnie McIver Brown, wife of Chadbourn's leading entrepreneur

Joseph Brown (see chapter 4), tried her hand at local-color fiction, she turned the familiar story of runaway slaves escaping to the Green Swamp, a fairly common occurrence prior to emancipation, into "A Romance of Slavery Days." The story's narrator, Uncle John Sutton, relates the tale of "Mike," a slave in Bladen County. Mike takes off for the Green Swamp's "famous semi-tropical jungle" and turns out to be perfectly at home there:

> He knew the secret places of the jungle, the inaccessible strongholds of privacy. He knew where the fish were plentiful . . . where the "pusley" (wild parsley) and the "poke salit" grew in the spring; where the crab apples, blackberries and huckleberries ripened in the summer; where the muscadines, persimmons, walnuts and hickory nuts could be gathered in the fall. He knew the haunts of the wild turkey and deer, of the opossum and coon, the squirrel and feathered game. Cattle too went into this swamp in the fall . . . so butter and milk, and even cheese were among the possibilities of menu.

Mike's woman Jane answers her mistress' probing about Mike's whereabouts with stereotypically Janus-faced disdain for his disappearance: "Den he's same as er dead nigger, kase dem bears in dat swamp gwine ter make quick wuck er him some uv dese dark nights. Dey do say de hants walk dat place." Then Jane escapes to the swamp to join Mike, gives birth to two children (Uncle John being one of them), and emerges from the swamp with her family when they learn from lumbermen that the Civil War has brought an end to slavery.[31]

The legends surrounding the original settlers on the rugged outpost called Crusoe Island, about six miles down the Waccamaw River from its mouth at Lake Waccamaw, further tie wilderness to outlawry, brigandage, and popular conceptions of race. One legend claims that the original settlers were Europeans and Native Americans who intermarried and were subsequently driven from the coast by whites outraged at this kind of race-mixing. Another story relates how pirates drove the all-white settlers from the coast, while a third legend claims they were themselves pirates who sought refuge in the interior. A fourth legend identifies them as renegades or deserters from either the Revolutionary War or the Civil War; yet another story holds that they were survivors of the "Lost Colony" of Roanoke (a story with its own ghastly dimensions). Finally, according to the legend that has received the most attention in the local and regional press, the little island was settled by Jean Formy-Duvall, a surgeon in the French army at the time of the French Revolution, and the group of royalists he reportedly saved from execution for treason. Formy-Duvall, whose name is still common in the area, feigned the condemned traitors' deaths and fled with them to Haiti after one of the men presumably dead was discovered to be alive. During the Haitian slave revolt of 1794, they left the Caribbean and made their way to the North Carolina coast before retreating into the interior.[32]

These legends and stories endow the wetland environment with a certain mystique—an air of danger, adventure, and mystery—that is peculiar to swamps and central to the popular mythology and general worldview of people living in their vicinity. They focus our attention on the experiential knowledge people had of their wild surroundings. First, they consistently depict the swamp as an unknown, haunted place, home to mysterious powers and spirits, and as a place of refuge for the dangerous, the lawless, and the hunted. Second, the people who are most at home in the untamed swamp are those who were traditionally considered to be inferior to "civilized" people, exactly because they seemed closer to nature and therefore more animal-like: conniving blacks who were also more likely to entertain superstitious beliefs, stealthy Indians, and rugged backwoods people. (Marauding pirates, too, were uncivilized rogues of dubious lineage in the popular imagination.) Third, these stories fired the imaginations of people who were well-respected members of their communities: they reflected and, in the telling, reinforced commonly held values and attitudes toward uncivilized places and people; at the same time, they allowed citizens of towns on the swamp's edge vicariously to bask in the depraved glory of a wilder past, and to see their ability to live close to such depravity reflected in the sheen of their achievements taming the wilderness.[33] Fourth, the swamp mystique is rooted in romanticized and mythical versions of the pre-European, colonial, and antebellum pasts that nonetheless draw on real events: the flourishing and passing of Indian cultures; the American, French, and Haitian revolutions; slavery; disease; the lumber industry; and the coming of the railroad. Such rootedness in the broad sweep of history gives the swamp mystique great continuity and purchase in shaping the historical identity of the region's inhabitants, even when they were determined to transform the actual landscape. Most of all, it forces us to acknowledge the mythological dimensions of a community's view of its own history and existence, no matter how contrary to supposedly objective observations of reality they may seem.

"A fit abode for ferocious wild beasts and disgusting reptiles"

Visitors to the Green Swamp, too, tended to be blind to its ecological complexity; and even if they noticed it, they lacked the knowledge and terminology to describe it. Even if travel accounts are more realistic than legend and myth, travelers to the Inner Coastal Plain Region nevertheless depicted the landscape they encountered in very similar terms: they, too, saw a wilderness that was dangerous and threatening—best avoided, sped through, or tamed by civilized people like themselves. Southeastern North Carolina, the Cape Fear River, and especially the port cities of Wilmington and Fayetteville were regular stops for travelers on their way from Virginia to South Carolina and Georgia. Norfolk, Savannah, and most of all Charleston, that "easy, affluent, aristocratic city . . . whose very atmosphere is fragrant of refinement," were obligatory destinations for genteel European visitors who desired to see the famed elegance of the South.[34] However, when they crossed by coach, on horseback, or by boat, the vast forests and swamps that lay between the coastal cities, these travelers usually reported getting an unwelcome taste of something rather different: the monotony of a landscape of endless pines.

To be sure, the dry, sandy areas that were richest in pure longleaf pine growth were the least harrowing to traverse, and observers often simply did not see the many places, plants, and animals that contributed to the region's biodiversity. Rare was the traveler who, like the visitor to the Cape Fear in 1734, spent time hiking and hunting in the dense woods that stretched between the various plantations.[35] For most first-time visitors, the defining characteristic of the South Atlantic pine belt was almost inevitably its "dreary desolateness."[36] Rough travel through the flat, sandy stretches of monotonous pine forest, although not as harrowing as stumbling through swamp thickets, only amplified their sense of the landscape's inhospitableness. And since most of them had to cross the pine barrens in order to get to the glamour of Charleston, it is not surprising that they experienced the longleaf pine forest as yet another discomfort of antebellum travel. Long, often interrupted train rides in uncomfortable cars; overnight stays, sometimes three or four to a bed, in the grungy hotels of frontier towns and dirty cities; and bumpy rides in packed stage coaches across rugged terrain—these did not sit well with people accustomed to a degree of luxury and leisure. "The road during the whole day lay through a miserable country," observed Margaret Hall in 1828 while traveling through the North Carolina tidewater: "pine barrens and swamps, with here and there a field of tobacco and cotton."[37] On her way to her new plantation home on the Cape Fear, Catherine Clitherall, about to be married, traveled a road "so narrow & thro' a pine forest, we . . . expressed to the driver a dread of 'wild beasts' emerging from the woods."[38] Similarly, the journey to Charleston after the wedding ceremony "was not very interesting, thro' pine woods, and miserable accommodation at night," and life in the South as a whole was rendered miserable and outright lethal by "the pernicious effects of the country malaria."[39] Even harsher was the indictment by the *New York Morning Herald*: "Scarcely would this Union feel the loss, if the barren sands and endless pine plains of the Carolinas, were sank into the ocean."[40]

At the core of this disdain for the "dreary" and "desolate" swamps and savannas of the coastal plain, however, lay shared attitudes toward nature that went much deeper than an incomplete view or a lack of comfort. Many eighteenth- and nineteenth-century westerners, upper-class whites in particular, tended to perceive the wilderness they encountered in the romantic terms of its "sublimity," the degree to which the landscape reflected what they believed was the awesome majesty and humbling power of its divine creator.[41] But while the natural world was the divine creator's gift to humans, their assigned role in it was yet one of dominion and transformation.[42] By the early nineteenth century, especially, they had come to believe that the specific ways God had created nature actually mapped out the direction of economic development.[43] To nineteenth-century Americans, internal improvements such as the building of the Erie Canal represented "a triumph of art over nature," by which they meant a triumph of human ingenuity over the obstacles the wilderness posed to the opportunities that were inherent in its creation.[44] Subduing the earth had come to mean *improving* it to further the progress of human civilization, be it through the excessive use of sunsets and rolling hills in landscape paintings, the construction of canals to connect two natural waterways, or the clearing of forestland for the cultivation of crops.[45]

And so dominion they claimed: although philosophers over the centuries have denounced western engagement with nature as incompatible with any more or less biblical notion of stewardship, westerners considered the cultivation and economic growth that was at the root of so much environmental change far less problematic. This was what they understood "dominion" to mean.[46] Confronted with the swamps and pine savannas of the coastal plain, "a fit abode only for ferocious wild beasts and disgusting reptiles," they sought to create pastoral plantations whose orderly and cultivated gardens amplified the disorder and frightful character of the woods around it and made their transformation all the more imperative.[47] Upon her arrival at the Hermitage at the end of her journey through the pine forest, Catherine Clitherall praised just such a landscape: The plantation's gardens "were extensive, & beautifully laid out. There were arbors, & summer houses at the termination of each walk—seats under trees, in the more shady recesses of the 'large garden.' Many evergreens, & a creek winding its course through the grounds."[48]

Reinforcing elite travelers' impressions of the unimproved landscape as a poor and barren wilderness was the apparent failure of its human inhabitants thus to improve the swamps and savannas. Wretched and degenerate in appearance, the "back-woods corn-crackers" and "vagabonds" that lived in and around the Green Swamp substantiated the belief that a "solitude of unsettled forest rapidly barbarizes" its sparse population. "The shadows of endless forests cow down man's ambition by their very vastness," warned a writer for *De Bow's Review*, further articulating the idea that wild nature corrupted civilized humans if humans did not civilize the wilderness first.[49] When sophisticated travelers encountered the people of the piney woods, the latter did little to undermine these perceptions. "From Petersburgh to Charleston we passed through only three small towns," wrote Adam Hodgson, who passed through Robeson County. "The manners of the lower classes . . . altered very sensibly for the worse. Their general demeanor became more rude and familiar, and their conversation more licentious and profane: their appearance also, was dirty, ragged, and uncomfortable."[50]

As with the legends and myths surrounding the Green Swamp, we should underestimate neither the endurance of these beliefs, nor the attitudes toward nature and people they engendered. More than two centuries after the colonial establishment of the Carolinas, Nathaniel Bishop's canoe trip from Quebec to the Gulf of Mexico took him through a Green Swamp region that, judging by earlier accounts, had barely changed. Paddling down the Waccamaw River, he crossed "a great and dismal swamp." The river's current was "tortuous, black, rolling." Water running upstream presented a "mystifying spectacle." "Festoons of gray Spanish moss hung from the weird limbs of monster trees, giving a funeral aspect to the gloomy forest," while the abundant mistletoe "gave a Druidical aspect to the woods." Diversion from the forbidding landscape came in the form of a representative of law and order: the boisterous, hard-drinking Jim Gore, whose ill-pronounced warning of the "reaches" (river bends) downstream as "wretches" made Bishop fear an attack of mean-spirited old women. But such fears were, of course, not ill-founded in these illegible interior regions that Bishop's friends at the Coast Survey Bureau had "not yet penetrated . . . with their theodolites, plane-tables, and telametre-rods."[51]

Half a century after Bishop's journey, travelers and sportsmen attested to the swamp's enduring mystique. John Martin Hammond urged fellow canoeing enthusiasts to check out the Lumbee River, one of whose great attractions was the Croatan Indian Reservation: "Romance is attached to the tribe by reason of the legend that they are descendants of the lost colony of Sir Walter Raleigh at Jamestown."[52] Another outdoorsman, Williams Haynes, depicted his canoe trip down the Lumbee as a journey to a different time and place; despite the landscape's far-reaching transformation that southeastern North Carolinians had effected since the late seventeenth century, the "almost virgin forest" was not only a "natural game preserve" but also a "tangled wilderness" where one "went eighty miles before [one] saw a sign of modern civilization." The few human inhabitants one encountered only augmented this sense of strangeness: there were the "*mysterious* blue-eyed race descended from Sir Walter Raleigh's Lost Colony," lumbermen who connived to call bottle-neck cypress trees "dram trees" so as to claim a drink of whiskey as they passed them, and "a *strange, suspicious* old man whose curt answers and strange advice sent our imagination weaving tales of moonshiners and blockade whisky stills."[53] Even Bland Simpson, who beheld much of "the *exotic* land of Maco, Crusoe, and Last Chance" with a far more ironic and ecologically sensitive eye, could not resist elevating the rare Venus flytrap to the swamp's main *mystery*: "not French swampers, not Whiteville conjurers, no voodoo or mojo, whether Haitian or homegrown, is more *mysterious* than the little green low-growing flytrap of Venus." But then again, by the time Simpson arrived, poachers of the small insectivore had come to be of far greater concern to law enforcement officials than "the swamp distillers of yore."[54]

"A lawless people filleth the lands"

Entrenched and enduring views of the Green Swamp region as a place of wonder, mystery, and danger constitute not so much a lack of proper ecological knowledge in need of correction as a mythological, experiential, and therefore different way of knowing nature—one that is not contradicted but born out by successful business development, and one that no amount of scientific discourse on "deep ecology" and biodiversity will easily displace. Already when the English began colonizing the Carolina wilderness, they brought with them these attitudes of fear and loathing toward wild places and all that lived within them. They also came with the determination, heightened by their violent colonization and subjugation of Ireland and the Irish, to transform the colony into a prosperous settlement that would benefit the English Crown. Since North Carolina was, as William Gordon informed his superiors in London, "a country but wild and imperfect in circumstances," representatives of the royal government and of various churches sought to make the land legible; the forests peaceful, productive, and orderly; and the people industrious, law-abiding, and religious.[55] Many of the documents they produced to describe both the colony as they encountered it and their own attempts to improve it resembled "cadastral maps": they made the land legible by describing it in the terms and categories of the emerging state's interests, by defining the proper understanding of what constituted a desirable and successful colony, and by identifying which aspects of colonial life—or which specific areas of the colony—required governmental intervention and adjustment. This, they hoped, would enable first the proprietors and then the royal government to impose their visions of proper land use on the Carolinas.

The most important objective the English government pursued was the extraction of wealth from nature by industrious and loyal colonists; many a report, therefore, while seemingly containing mere enumerations of flora, fauna, human inhabitants, colonial settlements, and economic activities, actually sought to indicate the many possibilities for further colonization and explanation. Between late September of 1663 and early February of 1664, Anthony Long, William Hilton, and Peter Fabian traveled as appointed commissioners from Barbados to the Cape Fear to explore the land on both sides of the river and report their findings on its usefulness to the English government: "The woodland . . . is, for the most part, pine, and in some places as barren as ever we saw land, but in other places good pasture ground." They came upon a sign put up by a group of New Englanders who appeared to have been sufficiently displeased with the area to leave a note informing anyone coming after them of their disdain. In response, the commissioners pointedly sought to impress upon their readers that the group's "disparagement of the land" and "great discouragement of all such who should hereafter come . . . to settle" constituted nothing short of "scandalous writing." They saw land that was "as well timbered as any we have seen in any part of the world, sufficient to accommodate thousands of our English nation."[56] An attempt in 1667 to recruit "ingenius and industrious persons" for a settlement on the Cape Fear similarly called attention to the area's wealth in timber, birds, and fish, as well as to the few settlers' success at growing cash crops.[57]

Works such as John Brickell's *The Natural History of North Carolina* (1737), which detailed the young colony's geography, flora, fauna, human settlements, and developing economy, satisfied the curiosity about faraway places that the Age of Discovery had generated. Abounding in what then qualified as objective, scientific observation, and organizing the bewildering natural world into a taxonomy of rational categories, they were, in another sense, typical products of the Enlightenment. Most important, though, they informed ambitious readers of the possibilities and opportunities for development and improvement they could find in the New World's as of yet "vast Tracts of Land lying waste, or inhabited only by wild Beasts."[58] Similarly, the author of a land survey informed the Lords Commissioners for Trade and Plantations (the Board of Trade) of the abundance of swamplands in the Green Swamp region, but only to assure them that they would be "the best Lands in the Province when reclaimed," especially the lands at the head of "Waggamaw Lake" where he lived. At a time when the boundary between North and South Carolina was a point of dispute, he also suggested that the Lords include the entire Waccamaw River within the boundaries of North Carolina for the purpose of trade, which would benefit both the colonists and the Crown.[59] These descriptions were indeed "maps" that told potential settlers not so much what there was to see as what they should look for once they were in North Carolina, what they should bring, and what kind of trade they should pursue in order to settle and thrive.

Those responsible for the colony's governance consequently supported settlers who used approved means to turn the colony's natural potential to their own, the proprietors,' and the Crown's benefit. The resulting expansion of the colonial economy's extractive practices into the Green Swamp region's illegible landscape was accompanied by attempts to render it legible according to these government and business dictates, and they reflected clearly the attitudes toward wild nature that informed the building of a mercantile empire. Not least among these attempts was a late-colonial policy that encouraged the manufacture of naval stores on American soil—a clear endorsement of extractive practices in the longleaf pine forest—while banning the colonies' production of other manufactured goods.[60] Many laws, similarly supportive of proper resource extraction, preceded it. The fundamental constitution of Carolina stipulated that from the outset of colonization "whale fishing . . . shall wholly belong to the Lords Proprietors." According to a 1673 law, nobody could "hunt, range for or kill wild or outlyinge cattle on any neck of land" without a "just claime" or proper permission.[61] In 1766 an act was passed "to Prevent the Unreasonable Distruction of Fish," which was considered necessary because "many Avaritious persons, by Extending Seans, Nets, Hedges, Stops or Dams . . . prevent fish from passing up . . . Streams, to the Prejudice of the Upper Inhabitants of the same." Avarice was thus considered improper when the destruction of resources it caused was "unjust," "unreasonable," and detrimental to the interests of other settlers; dams erected in order to operate mills were exempted, as these were essential to furthering just and reasonable economic development.[62]

Royal agents also expressed great concern about the many settlers who took up land as squatters and began improperly to use the forest's resources without legal permission or title to property. Not only did this infringe on the Crown's profits, it also prevented the erection of permanent structures that signaled the advance of civilization.[63] In 1738 a report to the Secretary of the Board of Trade claimed that unregulated use of natural resources was rampant, and tellingly defined the squatters' practices as illegal and wasteful, regardless of whether or not they might serve immediate, local needs: "It has . . . been a practice of long standing in the Colony for people to Box pine trees for Turpentine and burn light wood for Pitch and Tarr without taking out pattents for the Lands"; in doing so, these settlers had "made waste of the King's Lands."[64]

In spite of such outcries, many settlers in North Carolina continued to claim land, harvest or destroy timber, burn lightwood, and choose deliberately not to pursue their legal title once they got from the forest what they sought. In 1772 Governor Josiah Martin issued a proclamation forbidding those who had not fully cleared a well-defined land grant with the secretary's office to enter any lands and engage in extractive activities. Anyone entering ungranted lands to gather lightwood would also be prosecuted for their offense.[65] In this case, several people spoke out against the proclamation because they believed it would not benefit the Crown: the lands in question were not really of interest as they were of marginal quality and generally taken up for sawmills and ranging cattle. Moreover, enforcing the proclamation was expected to cause the squatters to pillage the land while they could or, even worse, turn to still vacant lands. After all, an opponent of the proclamation wrote to Martin, "In all probability the vacant poor Lands will be Pillaged by the destruction of the Timber and Lightwood and the Range destroyed . . . by Transient People who are continually passing from one Province to another." Experience had taught that squatters of this ilk resisted all manner of government intrusion and were "ready to fly from one Province to another to shelter themselves from prosecutions."[66] While enumerating local militias, for example, the government found "Waggomas" (Waccamaw) wanting. A little ways west, a settlement furnished only "a mixt crew": there at the head of the Little Pee Dee River, "a lawless people filleth the Lands without patent or paying quit rente." They even "Shot a Survey'r for coming to view" some lands in the swamp.[67] In the case of attempts to restrict access to mapped-out forestland, therefore, officials believed that intervention would only generate more undesirable disorder and transience.

The common practice of setting fire to the woods in order to clear land or improve its fertility also did great damage to property, and was all the more detestable when people used fire carelessly or irresponsibly. The phrasing of the law to prevent destruction by fire is once again very illuminative of the purposes it served: Fire was "destructive to Cattle and Hogs, extremely prejudicial to the Soil, and oftentimes of fatal Consequences to Planters and Farmers, by destroying their Fences and other Improvement." People were prohibited from setting fire to any woods except their own property, and then only if they properly warned their neighbors and took care to extinguish the fire before it did damage to lands adjacent to their own. Lawbreakers who could pay were to be fined ten pounds plus damages; slaves, free blacks, free mulattos, and vagrant persons unable to pay were to receive thirty-nine lashes on the back at the public whipping post.[68] Thus, although fire was a common and natural occurrence in the longleaf pine forest, one essential to its ecosystem, the law served to protect property and improvement against its "improper" use.

Another age-old human subsistence practice, hunting animals for food, had by the early modern era generally fallen under similar control by the government, which defined lawbreakers as deviants from the norms of order and proper use. Hunters targeted by an act "to Prevent the Pernicious Practice of Hunting with a Gun in the Night by Fire Light"—a technique some settlers apparently used to kill other settlers' horses and cattle while pretending to be hunting for deer—were described as "not disposed to pursue some honest employment in order to obtain a Subsistance."[69] A 1745 law served "to prevent killing Deer at Unreasonable Times, and for putting a stop to many Abuses committed by White Persons" who pretended to be hunting deer in the woods or on "unfenced Grounds." Once again, people who had properly fenced-in property were exempted; the law targeted "idle and disorderly Persons, who have no settled Habitation, nor visible Method of Supporting themselves, by Industry or Honest Calling, many of whom come from Neighboring colonies, without proper passes." These transgressors, the law claimed, killed deer out of season, allowed their carcasses in the woods to attract wolves and other vermin, and stole cattle and horses.[70] When the Act was amended in 1766, it was clear that the mobility and refuge the woods offered these "Disorderly and Dissolute Persons" made fines hard to collect: the offenders, "having no property of their own," were still "assembling in great Numbers, and camping in the Woods, and kill Deer, burn and destroy the Range, burn Fences and commit many other injuries."[71]

International relations presented another challenge to order and legibility, as did the kind of people who deliberately operated outside the official realm of diplomacy and thereby severely complicated matters of governance. The long battle between Europe's imperial powers for hegemony in the Americas exploded into violence in North Carolina's southeastern corner in 1747. That summer, Spanish ships landed on the Outer Banks, and their crews killed all the cattle and hogs of the so-called Bankers. In 1750 when a Spanish ship in need of shelter and repairs sought to dock in North Carolina, its crew feared retaliation from the Bankers, who were known to be "very Wild and ungovernable."[72] Privateers and pirates also threatened the establishment of an orderly society on the coast, further adding to the notion that the Cape Fear region was "no man's land." In the early eighteenth century, pirates set up shop at the mouth of the Cape Fear—at the current site of Southport—and preyed on the commerce of the North Carolina town of Charles Town until 1719 when an expedition from Charleston, South Carolina, ended the plundering. The legendary Blackbeard, too, had his headquarters in the Carolinas and enjoyed unofficial protection from Governor Charles Eden in exchange for a share of his loot. When Virginians defeated the notorious pirate in 1718, they accused North Carolina of granting the pirate and his crew a safe haven while the coast trembled in fear.[73] Little wonder that a representative of the British government in North Carolina, in a letter to Lord Cartwright, solicited royal approval for "the part I have lately acted in Rescueing the Trade of North Carolina from the Insults of Pyrates upon the Earnest Solicitations of the Inhabitants there."[74] Little wonder, too, that the colonial government in 1745 passed an act to encourage settlers to move to Brunswick and counter the threat of pirates and belligerent foreign ships with the order of a well-established, industrious town; nevertheless, in 1748 a small squadron of privateers from Havana sailed up the Cape Fear River and ransacked the town.[75]

When two of the communities along the Cape Fear turned the misfortune of these privateers—one of their ships blew up, and the North Carolinians managed to salvage some of the treasures and slaves on board—to their advantage by using proceeds from the sale of their property to build churches in Brunswick and Wilmington,[76] they pointed to yet another shortcoming of the southeastern coast: its lack of religion. Attempts to bring the Gospel to settlers there were of a kind with governmental attempts to punish the idle and transient and protect the industrious and sedentary, and they ran into similar obstacles. Churches popped up only intermittently in the sparsely populated region, and it was rough going for those who sought to carry the word of God further into the swamps.[77] A group of traveling Presbyterians found themselves spending the night in the Green Swamp "surrounded . . . by bears, wolves, and panthers." Only their campfire had, "under Providence," protected them "from their fury, and from being affrighted by their hideous howl." Yet the danger of wild beasts lurking in the swamp paled in comparison to the wickedness of the company they met "at a miserable inn." Under the canopy of heaven, God protected them from animals that generally attacked humans only "of necessity, or by provocation"; but it seemed that nothing could protect them from creatures "so depraved as to render them more terrible and dangerous," driven to attack by "the incitements of their depraved passions."[78]

Itinerant preacher Charles Woodmason tied this lack of holy living in North Carolina to its inhabitants' degenerate background: "The Manners of the North Carolinians in General, are Vile and Corrupt . . . And how can it be otherwise? The People are compos'd of the Out Casts of all the other Colonies who take Refuge there."[79] Nicholas Christian, a preacher in the town of Brunswick, reported his strenuous efforts to preach to communities around the Green Swamp in a letter that reveals how agents of civilization contributed to the swamp mystique:

> The roads [are] exceeding bad especially to Waccamaw there being upwards of twelve swamps to cross . . . In this place common prayer had never before been read nor the Gospel preached, so that the people are rather ignorant, but very desirous of being instructed which, according to the utmost of my ability, I shall endeavor to do . . . This place is situated on Waccamaw river near the green swamp entirely out of the way of all Society.[80]

Christian, like his fellow evangelizers and the royal official lauding his own efforts in the face of piracy, did make his efforts to spread the Gospel look all the more committed when he described in colorful detail the trying conditions under which he labored for the Lord, specifically mentioning the poverty of the Waccamaw parish to call attention to his own cramped financial situation. In a larger sense, the frightfulness of swamp and forest not only amplified the obvious need for evangelization and the praiseworthiness of the bringers of good news; it also reinforced the fascination with the wilder parts of God's creation as a fount of glory for those who dared to penetrate it for the sake of the Gospel.

Heathen native peoples who responded unfavorably to the relentless encroachment on their territories by Christian Englishmen and other Europeans were another pest that required governmental control. Of course, the presence of the Tuscarora Indians, of a Souian tribe called the Woccons, Waccomassees or Waccamas (to whom whites referred as Cape Fear Indians), and of the enigmatic Croatans or Lumbees indicated that the Green Swamp and environs were no virgin wilderness. Their settlements and economic activities reflected a long history of environmental disturbance and human adaptation. Hunting, gathering, agriculture, and especially the use of fire associated with subsistence practices had altered the landscape. The use of timber for the construction of permanent or semi-permanent villages translated into significant, man-made structures in the forest environment. Trade routes that stretched into the Southwest had introduced non-native seeds and species as well as non-local artifacts into the environment. When native tribes aided white settlers as guides and hunters, they also contributed to major declines and shifts in wildlife populations and distributions. Even when Indians moved on and allowed the woods to reclaim places they had cultivated, the forest was no longer "virgin." But none of this registered with colonists—to them, a seemingly untamed second or third growth was as wild as a "virgin" forest, while the presence of Indians, past or present, only contributed to its wildness. The Lumbee Indians who remained in the area after surviving the European onslaught of warfare and disease had, in addition, the reputation of being a mixed race, which only increased Europeans' belief that these natives were of a particularly mysterious and vile nature. Indian raids and warfare in response to violent European encroachment further threatened the stability of colonial settlements: the "unofficial" frontier skirmishes they often provoked between settlers and natives only constituted more disorder that was difficult to contain and, as had been the case in Virginia, could lead to even more severe frontier unrest. So even though the official policy of treaties and wars played into the hands of white settlers by steering the native population toward removal and near-extinction, the Crown's attempts at monopolizing the colonial response to the Indian problem also served to control potentially unruly colonists.[81]

In addition to the disorder natives and settlers created on the colonial frontier, nature itself could wreak havoc. Fires, storms, hurricanes, droughts, and floods were difficult to control, and humbling reminders of human impotence in the face of nature's destructive power. Wild animals, especially large predators, also caused great "prejudice" to settlers by destroying crops and livestock, as well as by occasionally threatening humans—these pests, however, settlers believed they could control. Several acts "for Destroying Vermin in this Province" served to entice settlers to eradicate wolves, bobcats, bears, and panthers, as well as squirrels and crows, by offering them rewards for each animal they killed. The thinking behind this eradication campaign reflected the larger ideology of proper improvement, as a letter from 1777 to the *North Carolina Gazette* makes clear. Its author proposed increasing the bounty for killing wolves to five or six pounds per wolf because the then-current rates were not enough to entice people to kill them:

> Every grown wolf does 20 or £30 damage yearly . . . The object of that expense would be decreasing yearly, & before many years it would be a rare thing to hear of a wolf &c. doing damage among our settlements, and in proportion as these noxious vermin are thinned and destroyed, our cattle, sheep, and hogs, would augment . . . The in-

crease of our live stocks, where no casualities [*sic*] obstructs, may be compared to compound interest.

"Success" and "value" were at stake when wild predators not only attacked the livestock whose proliferation constituted capital growth but also kept these four-legged agents of colonization off "waste lands."[82] Destruction of such pests, in this view of progress, was only reasonable, especially when settlers also attributed to them the same demonic qualities they detected in Native Americans.[83]

But once again some settlers only exacerbated the disorder wildlife caused when they abused these policies; already in 1754 the act to destroy vermin had to be amended because it had "induced several idle stroling [*sic*] Persons to hunt and range without the inhabited Part of this Government to Kill the said Vermin, thereby to entitle themselves to the said Rewards." The amendment stipulated that settlers could only claim their reward if the vermin had been killed within ten miles of an inhabitant's plantation, the killer swore an oath to the veracity of his claim, and a judge certified it.[84] Because it was impossible to verify where an animal had been killed, enforcement of the act was predicated on a system of trust, honor, and judicial discernment of character. There was a proper way to hunt predators, which made the animals' wildness reflect positively on the hunter. Perpetrators of felonious dishonesty, however, were identified as "idle" and "stroling": those who would commit such acts were lazy, transient, dishonest, dishonorable, and outside of the law, all of which put them at odds with the dominant system of improvement and governance.

"A very kind gentleman, a very religious house"

The English Crown and its representatives in North Carolina saw their efforts to manage the colony's natural riches both in general terms of "civilization" and in more specific terms of profitability, propriety, and legitimacy. In the process, they defined activities and attitudes that obstructed the colony's improvement as wasteful and destructive, and therefore invalid and illegitimate. The perceived benefits of this kind of government-sponsored, economic-environmental progress, as well as their unequal distribution, reflected the imperial imposition of an obviously contested set of values on complex webs of life and myriad systems of sustenance, interaction, and exchange. From the earliest days of white settlement, colonial planters and farmers, enabled and encouraged by land grants as well as laws that protected their private property and defined their power over their slaves as absolute, capitalized on the natural movement of water and the abundance of pine and other trees along the many oceanbound streams. Planters along the Cape Fear, like those along the rest of the southern Atlantic seaboard, used the tides and the harshly extracted labor of West African slaves to create rice fields in the marshes of the coastal plain. They let their livestock loose in the abundant forest, as free-ranging cattle and hogs saved them the costs of feeding and fencing. Along the coastline, along streams, and on lakes, settlers and their slaves used lines, nets, and weirs to catch the many species of fish, shrimp, eel, oyster, and turtle that provided excellent nutrition. Once early colonists discovered the high sap content of longleaf pines, they and their slaves began to tap the trees to extract turpentine and produce naval stores for the empire's rapidly expanding shipbuilding industry. Planters and farmers used the natural currents to ship rice, naval stores, and timber to port cities such as Wilmington and Georgetown, whence merchants shipped them to northern Atlantic ports, the Caribbean, South America, and England.[85]

The wresting of profitable systems of slavery-based naval stores production and rice cultivation from the wet and wooded frontier economy bore the characteristics of the larger circum-Caribbean economy: largely uninhibited—when deemed "proper"—exploitation of nature, a brutal "use-and-discard" approach to human labor, and the elevation of skilled workers to privileged, near indispensable slaves.[86] As was the case in the sugar plantation complex that made Bahia, Jamaica, and Barbados such prized colonial possessions, the creation of profitable rice fields in the Carolinas entailed not only the slashing and burning of forests, the clearing of swamps, and the constant importation of fresh slaves to replace workers who succumbed to sickness and death in the treacherous marshes, it also relied on highly skilled slaves to tend to the technology necessary to manipulate natural processes.[87] Once in operation, plantations such as those belonging to "King" Roger Moore, a wealthy and influential planter whose family made its way from Ireland to the town of Brunswick by way of Flanders and Barbados, consisted of an intricate network of canals and locks by way of which skilled slaves controlled tides to assure proper flooding and drainage of the fields.[88] The production of naval stores involved cutting "boxes" in the base of pine trees and scarifying the trunk to force the sap to run into these large cavities, or cutting v-shaped channels as high into a tree as a worker could reach and collecting the turpentine in a boxlike construction. After workers used dippers to transfer the turpentine to buckets or barrels, they distilled and boiled the sap to produce the primary products of the naval stores trade: rosin, pitch, and turpentine. Once trees had stopped yielding profitable amounts of sap, workers either cut them down or left them to drain and dry for a few years until they died and the wind blew them down. Pieces of this windblown lightwood, subjected to a slow fire in an earth-covered kiln, produced tar.[89] Slaves often managed to make themselves integral parts of these production processes as sugar burners, tidal workers, bird hunters, tree boxers, distillers, and tar kiln operators. In addition, colonists transported their plantation produce down rivers and streams in wooden barrels, enabling slaves to distinguish themselves as coopers, rafters, and river pilots.[90] The benefits derived from the growing slave population's grueling toil in swamp, field, stream, and forest not only affirmed the validity of these economic practices and the worldview that underpinned them, they also helped elevate otherwise rigidly controlled slaves, from the perspective of improvement and legibility, to a position of far greater value than the idle, transient "white trash" in the piney woods.

This burgeoning plantation complex, with its socioeconomic stratification, was firmly embedded in England's Atlantic economy: its development would not have been possible without the slave trade, the various markets for colonial produce, the legal framework, preexisting attitudes toward the unruly and the wild, and the legal, military, and naval protection the mother country provided. The solidly eastward gaze that characterized much of the Lower Cape Fear's society was as much a product of historical circumstances as of geographical realities, if not more so: the society that developed on the southeastern North Carolina's marshy coastal plain, and especially the economy that took root there, was a product of the developments that shaped much of the Atlantic world.[91] Like so many wild and imperfect places across the Americas that were absorbed into the hinterland of growing port cities, the Green Swamp and its larger environs increasingly were subject to the forces that turned these booming ports into the most important outposts of Britain's Atlantic empire.[92] After the demise of Brunswick, a few miles to its south on the Cape Fear River, Wilmington emerged as the primary conduit for the many products harvested along water routes and shipped downstream, as well as for luxury goods arriving from Europe and the West Indies. Because of its rise as a major port, Wilmington was home to the region's mercantile community, its shipping industry, and a host of entrepreneurs and laborers caught up in the commercial life of the city. Sailors, fishermen, dockworkers, slaves, slave traders, raft pilots, warehouse managers, naval stores inspectors, customs officials, preachers, and innkeepers were among the many souls who turned the east bank of the Cape Fear into a vibrant place of business and diversion. Settlers upstream depended on the merchants and workers of Wilmington for commercial services, be they the purchase, storage, and exportation of plantation produce, or the importation and sale of groceries and luxury goods that arrived on the many ships that docked in the harbor. Even merchants in Charleston, which served as a similar market for interior producers, sometimes called on their colleagues in Wilmington to fill large orders for naval stores or lumber bound for the West Indies and were willing to offer the Wilmington businesses "the very best of terms" for their expedient compliance.[93]

The city thus served as a primary agent of colonization: settlers pushed upstream into the backwoods to open up new resources for commercial exploitation while remaining within the orbit of the Wilmington market and the British Atlantic economy, which dictated the terms and value of proper land use, profitable labor, exploitable resources, and successful cultivation.[94] The settlements that sprang up around the Green Swamp were frontier outposts of British "civilization" that after the fashion of a cadastral map delineated the swamp as separate from the larger coastal plain ecosystem and cordoned it off as a distinct realm for proper exploitation. From the outset the colonization of the Inner Coastal Plain Region was characterized by tension between, on the one hand, those who stood to benefit from proper, just, and reasonable exploitation, and, on the other hand, those who engaged in practices that did not fit the official model of colonization, and that their critics therefore deemed unjust, unreasonable, lawless, and destructive.

By 1765, when John Bartram explored the Green Swamp, Lake Waccamaw, and the Cape Fear River, the landscape hosted the agents of colonization that would continue to shape environmental transformation. There was the plantation, with its productive fields and orchards, its well-controlled labor force that came to be seen as part of the planter's extended family, and its benevolent influence over its inferior, sometimes recalcitrant neighbors. Bartram spent the night at several such plantations and described one of his hosts as "civil," another as "a very kind gentleman." There was the city, the gateway to the Atlantic economy, with its commercial infrastructure and expanding reach: Charleston, Georgetown, and Wilmington were all points of departure and arrival for Bartram. There was the church, whose increased emphasis on individual, inward holiness posed no real obstacle to accumulativeness and economic development—even if elite Christians frowned on the emotional, rowdy worship of their slaves and their backwoods brethren, at least religion buttressed social stability and order. Bartram lodged at the "very religious house" of a New Light Baptist who performed "family duty evening & morning." There was business development in city and country: Bartram noted tar kilns, straight roads, inns, and the appearance of a new sawmill. And there was the state, whose as of yet incomplete control over renegade offenders against the dominant order nevertheless solidified their social status as disorderly, destructive, lawless, and subversive of profitable legibility. Even the science that Bartram brought to the area served the established order: his enumeration of plant and animals species, especially in terms of their potential economic value, constituted a taxonomy that ignored the intricate connections between human and non-human organisms and instead provided yet another map to guide improvement.[95]

In the eyes of government officials and the planters, farmers, and visitors who shared their outlook on propriety, unruly settlers took advantage of the illegibility of wild nature—and added to it—by moving around, squatting, carelessly using fire, burning tar and gathering wood on lands not their own, and hunting in ways detrimental to nature and society alike. If the environmental practices the agents of colonization did endorse—settling permanently and legally, cultivating the soil, harvesting trees, eradicating vermin, hunting lawfully—were at times equally invasive and often more permanently so, at least they produced for Crown and colony the material benefits that validated the motivations behind colonization. Not least among these was a legible landscape of cultivated, productive, and taxable fields—inhabited by settled cultivators and industrious, well-behaved slaves—that reflected governmental and elite efforts to inscribe on the Green Swamp region a capitalist worldview of economic progress, individual industry, improved nature, and a stable social order. By the time discontent with British control over this legible landscape ignited the Revolutionary War, it was not this worldview that was being contested, but rather who was to be in control of its continued imposition and continental expansion.

2

"Turp[t] Is the Word!"

Late in 1776 naturalist William Bartram continued his exploratory trip through the newly independent southern states—a journey that had already taken him as far south as 90 miles below St. Augustine and as far west as Mobile—by traveling up the Cape Fear River. While observing and recording the plant life and general landscape he encountered, he commented repeatedly on the economic activities with which settlers had transformed, or stood to transform, the natural environment. He observed "vast rich swamps, fit for the production of Rice"; "a road . . . not far from the saw-mills"; and promising forestlands. "When these lands are cleared of their timber and cultivated," he predicted, "they produce abundantly."[1] There was, of course, nothing new about human beings altering nature for economic purposes, nor was it unusual for travelers to comment on such activities. A decade earlier, Bartram's father John had done so, and William was following in his footsteps. New was that by the time of the younger Bartram's sojourn, the region was in the throes of a revolution that accelerated dramatically what Bartram and his contemporaries considered humans' dutiful efforts to subdue the earth and make it fruitful. The American Revolution, that is, unleashed a torrent of egalitarianism and economic activity in the new nation's forests, fields, and streams that rapidly advanced the taming and cultivation of wild nature within a developing capitalist economy. It sparked the migration of unprecedented numbers of people searching for land and economic opportunity in a society whose traditional structures of social stability were rapidly eroding.[2] The events of the American Revolution created in North Carolina's longleaf pine belt, as it did across the North American seaboard, socioeconomic circumstances conducive to the expansion of the existing infrastructure of colonization, which soon surpassed in magnitude and geographical reach the ties between riverside planters and coastal merchants. Although Bartram himself kept remarkably quiet about the events that consumed the seaboard in the mid-1770s, his observations on landscape transformation and Native American conflicts hint at the momentous changes unfolding before his eyes.[3] The various native and imported cultures that had long contested the so-called middle ground were giving way to an expanding Anglo-American culture that saw in a landscape of rice plantations, turpentine orchards, roads, sawmills, and productive farms a sign of successful environmental exploitation and economic growth.[4]

Observers with a sympathetic eye for the changes that were sweeping across the young republic saw in the cultivation and settlement of forest and swamp a transformation quite different from the barbarization of backwoods settlers that travelers in pursuit of civilized leisure tended to see. They recognized in the crude inroads settlers were making into the wilderness the taming of the coastal plain, signaling not only the transformation of rugged wilds into productive fields, but also the transformation of a society of colonial plantations strewn along the Cape Fear into a dynamic, expanding, enterprising democracy. As an unidentified traveler to Brunswick County in 1816 remarked, "Man had left his 'footprints' even in the wilderness." Brunswick in the early nineteenth century presented visitors with a mosaic landscape, a mixture of large stretches of open forest, dense swamps, meandering streams, a rugged coastline, farms, roads, primitive manufacturing efforts, and a few towns. "The face of the country is generally low and level," the unidentified traveler reported, "and the soil sandy, with a growth of pine trees; interspersed with cypress swamps and open savannahs." Most conspicuously wild was an "extensive dismal, called the green swamp," but waterways and public roads made travel and transportation between Smithville, Fayetteville, Wilmington, Georgetown, and Charleston "tolerable." Smithville, on the coast, was a town of three hundred, with sixty houses, a school, and a courthouse; planters on the Cape Fear produced a "considerable quantity" of rice; farmers let their "cattle and hogs run wild in the woods" and raised sweet potatoes, Indian corn, peas, and a little cotton on cultivated land.[5] In 1820 another traveler remarked on the "rapidity with which a settlement is occasionally effected" when he stopped at "a very comfortable log-house" in the forest: "Thirteen days previously this was the middle of a wood, and not a tree was cut down!"[6]

What observers witnessed in forest and swamp was the transformative power of human activity and ingenuity on a "frontier of change": the Green Swamp region was one of the many places where efforts to overcome nature's challenges inspired the technological and entrepreneurial advances that drove the rapid expansion of America's capitalist economy.[7] Faced with nature's great economic potential and the obstacles to economic growth it presented at the same time, the people of the Green Swamp region had already built relatively small communities that served as outposts of the larger Atlantic market to which Wilmington gave them access. When the American Revolution dislodged the southeastern corner as one the commercial nodes in the British imperial trading network, however, it opened the region up to the developments that thrust crudely capitalist economies across the country into a nationwide "market revolution."[8] Around the Green Swamp, tar and especially turpentine, already important enough crops during colonial days to serve as currency, became the products around which revolved population shifts, the rejuvenation of plantation slavery after a brief decline, the growth of towns, and the creation of a vibrant naval stores market through the improvement of domestic processing and transportation facilities.[9] "Turp[t] is the word!" exclaimed Wilmington's Robert W. Brown to New York merchant Oliver H. Jones in 1835 as "the present rage" for turpentine and other naval stores had driven prices as high as $3 per barrel.[10]

"One of the greatest gifts of God to man"

It appears that during and immediately after the American Revolution the close ties between the naval stores trade and the plantation economy came somewhat undone, and production shifted to small farms.[11] Travelers and merchants in southeastern North Carolina tended to observe small farmers, rather than planters and slaves, engaged in this activity.[12] Elkanah Watson, who referred to the forest dwellers he encountered in the midst of the Revolutionary War as "wild tar-burner[s]," noted that "the few inhabitants here and there in the forest, subsist by the chase, burning tar, and collecting turpentine."[13] In 1833 merchant Robert Brown notified Oliver Jones that he expected to see turpentine appearing in town "as country people get relieved from crops."[14] In an 1843 article on the naval stores trade, he again noted that "tar is rafted to Wilmington . . . and bought of the country people by the raft."[15] As late as 1855 Amelia Murray, traveling from Petersburg to Wilmington, took evidence of turpentine and tar extraction as a sign of "poor people making the wilderness a springing well, and the barren land rich."[16] With the supply of longleaf pines seemingly inexhaustible, production of crude turpentine and tar in small quantities did not require much labor or start-up capital, so almost any small farmer who wanted to supplement his income could box a few trees or gather some lightwood to build a tar-kiln. At least temporarily, the absence of a strong government allowed small producers the easy access to forestland that the colonial power structure had sought to restrict, and they took advantage of it.

Another indication of the plantation's temporary decline as the regional economy's dominant unit in the years following the Revolution was a decline in the use of slave labor. After decades of growth, between 1790 and 1800 the area's slave population decreased from 5,452 to 4,852. In Bladen County alone, which had the largest area of land that offered planters immediate access to the Cape Fear, the slave population fell by one-third, from 3,408 to 2,278. During the same decade, the region's overall population increased from 13,481 to 17,739 inhabitants (31.6 percent), while a combination of new migration to the area and a high birthrate among settlers prompted a 59 percent increase in the white population. This demographic shift suggests that larger planters were abandoning their old plantations along the river for fresher lands farther inland. Some remained in the southeastern corner: the combined slave population of Brunswick and Robeson grew as the eighteenth century drew to a close. Yet a sufficient number of planters left the region to contribute to slavery's overall decline. Others freed their slaves when, in the aftermath of the colonies' struggle against "enslavement" by Great Britain, doubts arose over the compatibility of slavery with American liberty[17]: the number of "other free" people had nearly doubled by 1800, by which time slavery was a far less prominent feature of the longleaf pine belt's economic and social landscape.[18]

What did initially dominate economic activity around the Green Swamp, as they did in many parts of the young republic, were efforts to increase nature's productivity by way of internal improvements, especially the building and improvement of roads and waterways in order to boost transportation. Like so many Americans in the early nineteenth century, settlers on the Lower Cape Fear saw the richness of the natural world as a divine blessing bestowed on man that nonetheless required the transformative power of human "art" to help it reach its fullest potential. Many agreed with testimony of God's beneficence such as F. P. Porcher's: he proclaimed the stately longleaf pine, "by far the most valuable of the pines for tar, pitch, turpentine, and timber," to be "one of the greatest gifts of God to man."[19] With this view of providence came the belief that humans could and should to a significant degree channel the powers of nature, especially water; that is, by devising systems to temper its destructiveness, increase its productivity, and thus fulfill the promise embedded in God's creation.

The most popular method of internal improvement in the early republic was canal building—the canal bug that brought on the building of the Erie Canal and caused widespread canal fever across the eastern United States also bit the entrepreneurs of the Green Swamp region. They predicted that connecting isolated localities to a major commercial artery with an artificial, more easily controlled transportation route would open up the economic potential of an expansive backcountry and direct it away from rival ports. In 1798 a visitor to Lumberton noted that the state legislature intended to support the construction between Lumberton and the Cape Fear River of a fifteen-mile canal, which "would not only increase the quantity of produce in the county of Robeson," but "command that of most of the counties in the state, and a large proportion of that of South Carolina; the whole of which would center at Wilmington."[20] In 1816 the state government did indeed charter the Lumber River Canal Company, authorizing the State Treasurer to subscribe to $20,000 worth of company stock.[21] The project, like some earlier plans, never came off the ground, but North Carolina's representatives had shown their desire for the creation of man-made structures that could unlock nature's riches.

A second method was the improvement of existing waterways. The Cape Fear, after all, was the kind of river that could hinder transportation as much as it could facilitate it, especially upstream from Wilmington. In 1835, Robert Brown expressed hope that the lure of high prices would "hurry the country folks so that when there is water for rafts we may buy at same prices or others!" Obviously, the low river had drastically slowed down supply.[22] Snags, falls, and shifting shoals could also make trips to and from Wilmington lengthy and hazardous ordeals. The Cape Fear owed its name to the frightful experiences of seafarers attempting to enter the river across its dangerous shoals, and its wiliness was not limited to stranding and sinking ships.[23] When heavy rainstorms flooded the river, its raging waters swept away bridges and water-powered mills. As a result, businessmen were often reluctant to send larger ships upriver to Fayetteville and beyond, and those who depended on the river for a steady traffic in goods called for improvement of its navigability.

The state legislature had already responded to such calls in 1811 when it chartered the Cape Fear Navigation Company in order to improve navigability from the mouth of Cross Creek, near Fayetteville, down to Wilmington. Francis Waldo, editor of the *Carolina Observer*, was confident that the company would "render Fayetteville, as it was by nature intended to be, one of the most important towns in the Union." Given the inexhaustibility of the city's backcountry resources, he argued, "We may safely calculate upon its more extensive growth and prosperity."[24] People on the Lower Cape Fear banked on the company's ability, and on that of its successor—the Cape Fear and Deep River Navigation Company—to seize on nature's promise of abundance. Fayetteville, 100 miles from the ocean, felt justified to claim "all the advantages of a sea-port town" after "the valuable improvements in navigation."[25] By 1826, the *Wilmington Herald* thought it discerned a revival of its own city's "spirit of enterprize" and "extensive intercourse" with the Caribbean, which once again sent "the trade of the West Indies . . . into its wonted channel."[26]

As Wilmington thus regained its stature as North Carolina's "commercial emporium,"[27] the city's backcountry underwent the same kind of transformation that drew the expanding hinterlands of cities like New York, Philadelphia, Boston, and later Chicago into the orbit of urban-based capitalism.[28] The expansion of river towns such as Lumberton, Fayetteville, and Elizabethtown around the time of the American Revolution already indicated that market-oriented naval stores producers were a growing presence in the pine forest. Whereas large planters along the main waterways could ship produce to market directly from their riverside plantations, smaller farmers, especially those living farthest from navigable streams, depended on emerging small town business communities to handle their commerce. Elizabethtown, located along the Cape Fear, was Bladen County's only commercial shipping point. To get their products to market, Bladen County's farmers made barrels of turpentine and tar into horse-drawn land-rafts by running iron rods through them and coupling them in such a way that they could roll over the roads leading to town.[29] Lumberton, in Robeson County, served as the shipping point for cargo on the Lumber River. The same visitor to Lumberton who called for the building of a canal noticed a number of mercantile establishments "where large quantities of merchandize is annually vended in exchange for the commodities of the country . . . best suited to the European and West-India markets." Lumberton, he observed, "commands the produce of an extensive back country . . . its vicinity abounds with Lumber and Naval Stores, which are carried by water to George-Town and Charleston."[30]

These river towns came to operate as agents of colonization for particular localities in much the same way Wilmington served the Lower Cape Fear Valley as a whole. They facilitated the development of largely unsettled areas by providing the means by which the settlement and development of the wild backcountry could be effected. After all, for entrepreneurial spirits willing to face the heat and humidity of savanna and swamp, the southeastern corner was still as much a frontier of untapped opportunity as upstate New York, the Ohio Territory, Kentucky, or Tennessee.[31] In 1790 the area was home to 13,481 people, or fewer than four inhabitants per square mile.[32] It was not until 1820 that a total of 24,872 people inhabited the four counties (an average of 6.7 per square mile) and the "frontier" had officially closed.[33] Even then, much of the area's forests and swamps signaled opportunity rather than settlement, potential rather than achievement. But the growth of the river towns heralded major changes. Utilizing the town's geographically concentrated array of commercial services, a growing number of settlers capitalized on nature's resources and began pushing for further improvements that would turn larger and larger stretches of wilderness into a productive forest. As a result, river towns grew into the nodal points of a developing network of resource exploitation whose many branches stretched farther and farther inland while continuing to converge on the Atlantic coast.[34]

Before Americans managed to establish complete mastery over the waters that connected their settlements to markets, the railroad supplanted the canal as the main mode of transportation. As an engineer on the Wilmington & Manchester put it to a canal builder, "It may be true that you are gaining much information on canalling, but my dear fellow, the days of Canals are past."[35] In North Carolina's southeastern corner trains effected freer intercourse with the seaboard than had boats. Once large-scale production of rails and steam engines became a possibility and then a reality, railroad fever spread across America at an even faster pace than the canal craze. "Thus, will the product of the coast . . . find a market in the interior," a South Carolinian in 1832 described the inevitable benefits of a railroad from the Pee Dee River to the Roanoke in words reminiscent of those canal boosters had used only a few years earlier.[36] Several proposals for other railroads and charters for the corporations that promised to build them soon followed, and once again the dreaming up of plans proved far easier than their implementation in a sparsely populated area like the Lower Cape Fear region.[37] An engineer involved in the construction of the Wilmington and Manchester Rail Road described the atmosphere that surrounded the project as "rather foggy . . . we are going ahead on devilish little money." Especially troublesome was finding the necessary funds to lay track in underdeveloped areas such as Columbus County, a place the engineer deemed "a good definition of the word *poverty*."[38] This time, however, the people of the Lower Cape Fear and their representatives came through: by 1850, the Wilmington and Manchester connected North Carolina with South Carolina and Georgia; the Carolina Central Railway tied eastern North Carolina to the Piedmont; and the Wilmington and Weldon, at one point the longest railroad in the world, covered the length of the state's longleaf pine belt. Although the existence of railroad companies was often as precarious as that of the earlier canal companies, they did construct miles and miles of track that connected Wilmington to its growing and developing hinterland.

"Nature and commerce will compel full recognition of all we want"

Railroads succeeded where canals had failed because the successful development of rail transportation came at a time when technological developments in the production of naval stores sparked a boom in the naval stores trade and stimulated demand for reliable and expansive transportation facilities.[39] The discovery during the 1830s of a variety of new uses for rosin and spirits of turpentine—including their use as a solvent for crude rubber, as lamp fuel, and as a possible remedy for "colds, hoarseness, coughs, &c" when inhaled from a coffee pot—stimulated demand for naval stores,[40] while technological advancement in the form of an improved copper still finally made the domestic distillation of crude turpentine profitable enough for local entrepreneurs to boost turpentine production and set up in-state distilleries. In 1840 when Bladen, Brunswick, Columbus, and Robeson produced a combined total of 25,414 barrels of tar, pitch, turpentine, and rosin, only Columbus County listed "men employed" in the forest products industry. By 1860, however, the four counties were home to 267 known establishments (other than plantations) that employed 556 men in the production of crude turpentine, while 310 turpentine distilleries employed another 761 men.[41] With railroads passing through vast stretches of turpentine-rich longleaf pine forest and converging on a concentration of distilleries and auxiliary commercial services, naval stores became an attractive commodity for enterprising individuals willing to move to the Wilmington area in order to be close to the center of the trade. After increasing by only 12.2 percent between 1820 and 1830 and actually decreasing by 1.1 percent over the next decade, the population of the southeastern corner increased by 29.6 percent between 1840 and 1850 and by another 24.4 percent in the last decade before the Civil War.[42] By 1860 the many farmers among them had improved 218,288 acres of land for crop cultivation, less than 10 percent of the region's total acreage but still a significant transformation of forest into farm. In addition, 44,251 cattle, 23,394 sheep, and 108,753 swine roamed the region's unimproved and forest acreage, providing a significant source of sustenance and farm capital.[43]

Between 1840 and 1860, during which years advances in processing and transportation technology combined with a rise in demand for naval stores to turn North Carolina's longleaf pine belt into a seedbed for economic opportunity, larger planters also began to move their slaves to the pristine stands of the southeastern corner. As happened in neighboring Cumberland County, whose sudden population growth the *Southern Cultivator* ascribed to "about 300 whites and 700 slaves having arrived . . . to engage in the turpentine business," the growing naval stores industry made slavery once again one of the most prominent features of the socioeconomic landscape.[44] Whereas the increase in the overall population between 1790 and 1800 had been accompanied by a decline in the number of slaves, the increase between 1840 and 1850 was partly the result of a 42.4 percent growth of the slave population. In fact, within two decades the region's slave population grew by 77.6 percent, from 9,503 to 16,876. At the same time, no significant expansion of the cash crops whose rising values drove migration elsewhere accompanied this growth of the slave labor force, while the output of staple crops that did expand between 1840 and 1850 fell again during the following decade. In Columbus County, for example, a 63.9 percent growth in the slave population between 1850 and 1860 concurred with a 16.4 percent decrease in tobacco production and only a marginal increase in cotton output. Rice production did increase to a significant degree, especially in Brunswick and Columbus counties; but the fact that in 1860 Brunswick, with 3,631 slaves, produced 6.8 million pounds of rice while Bladen, Columbus, and Robeson produced only 270,893 pounds with 13,245 slaves suggests that rice prospects did not drive migration to the southeastern corner as much as did the turpentine boom.[45]

The growing population took full advantage of the navigable streams and the new railroads, enhancing Wilmington's position as the lower Atlantic coast's main commercial center and helping North Carolina shed its "Rip Van Winkle" image.[46] The editor of the *Wilmington Chronicle* chastised skeptics who had doubted the viability of railroad construction "whilst water facilities existed," challenging them to "visit the Depot, and see the freight trains coming in, laden generally, with from two hundred, to four hundred barrels of Naval Stores." And this despite the fact that freight rates were higher for the railroad than for water transportation—given "the damage to, and loss of barrels, the delays, the exposure of the rafts-men" that afflicted water transports, the railroad was an easy choice. Even farmers who had previously rafted their produce to New Bern now sought a market in railroad-supplied Wilmington.[47] Once the Wilmington and Raleigh commenced operations in 1839, suppliers kept up a steady stream of naval stores to Wilmington: 524 barrels of turpentine and 172 barrels of tar came in during the week ending on 1 May 1839, and 2,635 barrels of turpentine arrived during the five subsequent weeks.[48] When the Wilmington and Weldon Rail Road opened in March of the following year, supply increased even further: Wilmington agents frequently reported the arrival in the city of 1,500 to 10,000 barrels over the course of one or two weeks. So pleased were the Wilmington area's inhabitants with this successful cultivation of the longleaf pine forest that by the time they found Union ships outside their Confederate port they predicted the futility of a blockade in the face of so much human ingenuity. "We have . . . naval stores . . . even more exclusively than we have cotton," wrote the editor of the *Wilmington Journal*, who believed that the longleaf pine, without whose natural riches "navies are but a name," made a much prouder symbol of southern resilience and independence than South Carolina's "cabbage-headed" palmetto. "They may blockade us for a while," he wrote, "but in doing so they will be cutting their own throats. A little time and nature and commerce will *compel* a full recognition of all we want."[49]

"Runaways of bad and daring character"

If unimproved nature in and around the Green Swamp provoked the ire of many travelers, the growing impact of human striving on the environment only drew occasional criticism. Turpentine extraction, for example, although it left many trees standing for quite a long time, provoked Amelia Murray's disgust as she beheld the landscape's altered state during her travels in 1855. She minced no words as she described the unappealing sight of large stretches of forest "disfigured" and bound to be "killed by the bark being stripped off, that the turpentine may drip from it into a small vessel placed on the ground."[50] But the sporadic nature of such criticism, and its failure to inspire attempts to stem the tide of destruction, shows that only a few people entertained the notion that nature and commerce might somehow be at odds. The more than 100,000 destructive, non-native hogs that trampled undergrowth and ate longleaf pine seeds and seedlings barely registered when observers of improvement took in the landscape, despite later eliciting harsh words from many late-nineteenth century agricultural reformers and twentieth-century environmentalists. Of far greater concern to those who saw in turpentine extraction, railroads, and hogs the signs of nature and commerce working together to improve the wilderness were those forces and creatures that threatened to disrupt, corrupt, and sabotage the implementation of a proper, legible order on the landscape. The ascendancy of a technologically advanced, plantation-based turpentine industry at the expense of small, inefficient, occasional producers—not to mention of nature itself—was the result, not of the inevitable triumph of market forces that reward capital investment in efficient, large-scale production units, but of concerted efforts, backed by laws, government enforcement, and social networks, to impose on the longleaf landscape the legibility that such extensive and profit-driven exploitation demanded.

First, nature remained as tempestuous a mistress as during colonial days, often giving as good as it got. As inhabitants of a swampy coastal river valley, the people around the Green Swamp were very aware of the disruptive and destructive powers of nature. Not only did lightning kill people, it also caused destructive fires, exacerbating the damage done by storms as flames spread rapidly along waterfronts when warehouses full of flammable naval stores caught on fire.[51] In addition to the wiles of the main waterways, gales also continued to disrupt maritime commerce by taking down structures along the coast, blowing trees into the river, destroying crops, and upsetting traffic on rivers and ocean. The uncertain condition of the weather and of the ships that had to brave it caused businessmen great concern. The unfortunate journey of the frigate *Frothingham* from Wilmington to Martinique with a cargo of lumber and naval stores illustrates the risk of commercial seafaring. Having set sail on 18 July 1842, the ship hit bad weather in early August and endured three days of "frequent squalls and a very heavy sea," causing extensive enough leaking for the crew to keep at least one pump working and throw more than 10,000 feet of lumber overboard. After the storm subsided and the crew sailed on through moderate weather for a week and a half, the ship was hit with another gale; further leakage, the loss of sails, and a shortage of water forced Captain Henry Adams to anchor in the harbor of St. Thomas. After repairs, Adams steered the *Frothingham* toward Martinique on 30 September, only to have "squally conditions" throw it almost immediately against the rocks at the harbor's entrance. The ship was then condemned and sold in lots.[52] Given such uncertainty, Enoch Rust, a Wilmington merchant, informed the mercantile firm of William Walter & Company in Boston that he would write them about insurance for the *Commerce*, a ship bound for Jamaica, "as soon as I can ascertain whether she is safe in the river, as we have experienced a long and heavy gale from the N.E."[53] Sometimes Rust insured ships, but other times he acknowledged that man could never fully control nature, choosing to "trust in God and save the premium" instead.[54]

In addition to these less tractable forces of nature, wild animals continued to be a cause for disdain and alarm. The annual massacre of bobolinks—those "little destructives"—to protect the rice crop from their depredations contributed to the general decline of wildlife populations that manifested itself most clearly in the deliberate extermination of wolves, bears, and snakes.[55] The new North Carolina state government had predictably kept in place or amended the colonial laws that offered rewards for the killing of vermin.[56] Given the *Carolina Observer*'s initially serious response to the sightings of sea-serpents close to the Carolina coast (it became a joke when the truth was discovered), it is not surprising that fear, suspicion, and superstition continued to drive the extermination of "vicious" and "dangerous" creatures.

Second, as more and more people pushed further inland, they, too, became harder to track and control. In response, the state legislature kept in place or amended the various colonial laws defining the legitimate and illegitimate use of nature, as well as the appropriate punishment for violations. In the case of the legal prohibition on burning the woods, for example, the legislature increased the fines because it considered the earlier penalties insufficient to promote "the good purposes therein mentioned"; meanwhile, burning the woods continued unabated. A 1784 law specified the various hunting practices with which disorderly people undermined the proper exploitation of the state's natural resources. Hunting by firelight continued to be a problem, and the law specifically identified masters as the responsible party if their slaves used that particular method; killing deer and leaving the carcasses in the woods as unintended attractions for "vermin" remained an offense; killing deer between 20 February and 15 August unless hunters killed the animals on their own land was an unacceptable threat to the deer population; and hunting with a gun or dogs on other people's lands without permission was a clear violation of private property rights.[57] Similarly, in 1824 four campers found themselves turned in to the police for taking some wood from fenced-in private land: "The owners *lawed us*," they warned their fellow "Back Country Wagoners" of their experiences with the law's attempt to restrict what for generations of forest users had been a traditional, if contested, way of gathering firewood.[58]

Third, despite the march of progress and improvement in and around the Green Swamp, the land was still rugged enough to provide slaves with excellent cover for hunting, trapping, and gathering for unsupervised religious worship, and, most troublesome, for hiding. Work in the turpentine orchards was hard, and harshly extracted: "In the month of November, and through the winter season," escaped slave Thomas Jones remembered bitterly, "The men and women worked in the fields, clearing up new land, chopping and burning bushes, burning tar kilns, and digging ditches."[59] Punishment for the disobedient could be outright cruel, as is evident from the pain Moses Roper's master inflicted: "The first thing he did was to pour some tar on my head, then rubbed it all over my face, took a torch, with pitch on, and set it on fire."[60] Runaway slaves looking to escape, temporarily or permanently, the brutality of the regime fled to the forests and swamps and became another source of great anxiety among planters and farmers. In July of 1824 a slave patrol in Bladen County engaged in a violent skirmish with runaway slaves that left a slave and a patrol member dead.[61] In the 1850s, David Cecelski points out, "Runaway slaves . . . built at least eleven cabins and carved out a garden and grazing area in the midst of" the Green Swamp.[62] The maroon community managed to fight off a party of white raiders and kill the son of Richard Lewis of Bladen County, whereupon the senior Lewis petitioned Governor Thomas Bragg for aid in chasing these outlaws out of the swamp and bringing them to justice. "For many years past, and at this time," Lewis wrote, "there are several runaways of bad and daring character—destructive of all kinds of stock and Dangerous to all persons living by or near said swamp." The size of the swamp, he asserted, made it impossible for individual efforts "to dislodge these runaways, many of them belonging to persons unknown." And if running away (a form of stealing private property), destroying livestock (the destruction of private property), threatening and killing whites (the ultimate crime of racial insubordination), and belonging to unknown planters (legibility and order demanded knowledge of who belonged to whom) were not enough to insult the established order, the marooned blacks were so disrespectful of white superiority that upon killing David Lewis they "ran off cursing and swearing," yelling at the raiders "to come on, they were ready for them again."[63]

The growth of the turpentine industry also sparked anxiety about slaves employed on rafts, on boats, and in port towns. When planter Thomas David Smith McDowell of Bladen County informed some Wilmington merchants of "several thousand . . . staves belonging to some of the negroes" on board a raft coming down the river, instructing them to "sell them and send proceeds with mine and I will divide it among them," he described a practice common in the turpentine trade.[64] Yet the relative lack of supervision over slaves engaged in the maritime trades, as well as the relatively free mingling in an urban setting of slaves with free blacks, free mulattos, and the sailors and slaves that worked on ships from all over the Atlantic created a loose and fluid zone of interaction and activity that many slaveowners eyed with suspicion and concern. Already in 1785 North Carolina had passed a law aimed at clarifying and shoring up the policing of slaves in Wilmington and other towns, specifically their practice of hiring themselves out without written permission, as well as their interaction with free blacks and free persons of mixed color.[65]

Fourth, although the so-called golden age of piracy had long passed, pirates and other outlaws continued to cause concern among the inhabitants of coastal and forest settlements that offered so many quiet spots and excellent hideouts. In June of 1820, with the American victory over the pirates of the Barbary Coast in the recent past, the *Raleigh Minerva* reported the apprehension of pirates at Smithville; having sailed on a brig from Buenos Aires, they were apprehended off the Carolina coast on a ship they had taken at Cape Palos. The newspaper also reported that prior to the arrest of the pirates, eighteen of the crew had gone on shore at Lockwoods Folly and disappeared into the woods.[66] The connection between wilderness and outlawry also came to the fore in the dense woods of Robeson County during and after the Civil War: the Lumbee Indians of the Lowry gang, whom the law defined as "free persons of color," hid in the swampy woods to escape from conscription into Confederate labor service. Engaging in guerrilla warfare against soldiers and planters, while depending on communal support and traditional subsistence practices for survival, the renegade Lumbees often eluded capture or escaped from prison like the stealthy, slippery Indians of frontier mythology.[67]

"The manner of preparing it for market"

It was in this context of social, economic, and political flux, born of the American Revolution and reverberating through much of the nineteenth century, that well-connected planters and merchants restored to the plantation regime the hegemony it had enjoyed during colonial days. This time, their prominence was not the result of close ties to imperial seats of power and royal monopolies. Rather, it was the elite's success in anticipating, and responding to, the demands of large-scale, industrialized production and trade—demands in terms of technology, capital, labor, and business savvy—that eroded the position of smaller producers in the woods. As they projected a system of orderly, lawful, productive, and profitable resource extraction onto the land, planters, merchants, and other general entrepreneurs to a large extent came to own or control the trees, the tools of extraction, the means of transportation, the capital to produce naval stores, and much of the labor force. Joining forces in order to maximize profits from the tapping of trees, they turned developing networks of transportation, information, and capital into near-exclusive domains of plantation production and urban commerce. This left small farmers to play an increasingly marginal role as independent producers or a new role as wage laborers in the growing processing industry.

While we give the political radicalism of the American Revolution—or at least the promise thereof—its proper dues, we often ignore the more conservative nature of the economic order it made possible. The hustling and bustling, the innovation, the migration to new communities, the democratization, the widespread claims to freedom and opportunity, and the attendant erosion of the traditional social order caused much uncertainty, anxiety, and fear among existing elites and community leaders, soon generating a countermovement that sought not so much to reverse the transformation as to contain and direct it.[68] As a result, a process of political retrenchment at the federal level also manifested itself in so many ways, in so many localities, on so many levels.[69] Within the forest communities that sprang up and developed around the Green Swamp, planters and merchants used their economic and political clout to set the terms for economic and communal growth, often with the aid of fellow planters and like-minded farmers and merchants in the state legislature. As a result, an emerging business community of general entrepreneurs took it upon itself to dictate the terms of proper economic development through the definition and implementation of business standards and laws that governed pricing, inspection, and quality control. In the process they defined the practices of smaller farmers, "wild tar-burners," Native Americans, slaves, and others who could not be absorbed into this emerging business eco-culture, as illegitimate and destructive.[70]

The planter class's verbal, political, material, and financial support for internal improvements and elite leadership in southeastern North Carolina's longleaf pine forest economy stimulated the growth of an advanced throughput economy that, if not the pinnacle of industrial capitalism, was nonetheless essential to its development in the United States.[71] Advocates of internal improvements pushed for the construction of canals and railroads, not in response to an existing yet unfulfilled demand, but in order to create demand for those goods and materials for which the Wilmington hinterland could provide the natural resources. The Commissioners of the Cape Fear and Yadkin Rail Road suggested just that in a report regarding their projected railroad. Its construction would not only increase supply to Wilmington, they also expected that "a far better market for naval stores will be created, . . . the labour of hands and teams will be rendered more profitable, and a large amount of funds diffused in the neighborhood."[72] The development of advanced transportation networks, therefore, constituted a departure from earlier economic systems; apart from its successful breaching of the physical barriers between localities that, once bound together, could evolve into regional and national networks of commerce, it helped overcome the less tangible limits on the demand necessary to sustain the greater volume of goods now traversing the expanded routes of trade and justify production costs.

The transformative influence of the railroad worked on at least two intertwined levels.[73] Within the narrow realm of the naval stores trade, it helped stimulate enough demand to sustain a drastic rise in production. Within the broader realm of economic activity in a society increasingly devoted to commerce and industry, it stimulated all kinds of commercial exchange that translated into an increased, self-fertilizing demand for a range of commercial products and services. Market reports, a staple of antebellum newspapers, indicate that the railroad had a stabilizing effect on the price of tar and turpentine in a market where fluctuations in supply and demand typically caused prices to rise and fall. Before the advent of railroad transportation, prices gradually dropped as a result of increased production on the part of a growing population and slackening demand on the part of a more or less stagnant business. Turpentine gradually went from $4 per barrel in April of 1818 to $2.75 per barrel in September of 1823 and to $1.70 per barrel in February of 1831, rising intermittently when exchange activity drove prices up. Between 19 May and 17 November 1837, the price of turpentine rose from $1.25 to $2.70 per barrel, only to be back down to $1.50 on 9 March 1838. Given the sudden increase in supply that followed the coming of the railroad, as well as the boom in demand as reflected by the growth of the distilling industry and the number of exports from Wilmington, prices should have undergone even wilder fluctuations whenever increased supply preceded increased demand or vice versa. Instead, frequent turpentine transactions that involved 1,200 or even 1,800 barrels kept prices fairly steady. After an initial, gradual drop from $2.50 in June of 1839 to $1.60 in July of 1840, prices went back up and remained between $2.10 and $2.65 for all of 1842 with the exception of a little dip to $1.90 at the end of the year. The market saw neither steep drops nor spectacular climbs: the economic optimism that followed railroad construction generated enough additional demand for naval stores to offset the potential price-depressing influence of a large increase in supply.

The national boom in commercial activity that accompanied the transportation revolution generated technological advances and a demand for ships and shipping that furthered the naval stores trade as it developed into a predominantly plantation-based enterprise. The establishment of commission agencies for the express purpose of finding buyers for naval stores, as well as the construction of warehouses on the part of merchants confident enough to store unsold naval stores rather than accept lower prices, helped increase demand and temper potential price drops. The successful pursuit of economic growth created a belief in the possibility of further development, convincing people to engage in the production and purchase of throughput products. Price fluctuations of only a few cents in a market that hosted both domestic distilleries and ships bound for the Caribbean, New York City, Providence, Philadelphia, and Liverpool meant that producers of naval stores could generally count on cash income when the new railroads took their produce to the Wilmington market. But because of its increasing size and complexity, it was a market that favored those who could devote the necessary labor power to large-scale production, those who could purchase and operate the technology required for high-end production, those who could afford the railroad costs and administrative fees, and those who could read market reports and correspond with merchants and agents.

In conjunction with the people in the New York market, merchants in Wilmington and other coastal cities in the longleaf pine belt developed a naval stores grading and processing system that reflected both the increasing complexity of the market and the growing advantage of planters over small producers. Anyone shipping naval stores to the coast was charged for lighterage, freight, storage, weighing, and cooperage, as well as commission, insurance, and taxes. Produce was further subject to inspection and classification: by 1877 the mercantile firm of Williams & Murchison classified rosin as "strained," "good strained," or "fine"; crude turpentine as "hard," "soft," or "virgin." The market in New York went even further, as it quoted prices for twelve grades of rosin, including "No. 2," "Good No. 2," "Low No. 1" and "No. 1" on the lower end, and "Low Pale," "Pale," "Extra Pale," and "Window Glass" on the higher-priced end. It even quoted different prices for used barrels, with "N.Y. second hand" selling for $1.80 to $1.85 and "Southern second hand" doing only $1.55 on 8 August 1877.[74] As a result of these changes, the "wild tar-burner" found himself on the bottom rung of an increasingly stratified, complex system of exchange. Lacking the proper technology, he could only expect to produce lower grades of tar and rosin, which sold for the lowest prices. This was especially so after the enactment of "more stringent inspection laws" designed to battle the low quality of tar polluted with sand, "a pint of which will condemn a gallon of tar."[75] With all the added costs, the small producer had little hope of making a profit that was worth the effort or of understanding the larger market in which he played an increasingly smaller part.

Further widening the gap between large and small producers, entrepreneurial planters shared their expertise in an industrial forum that emerged within the South's agricultural literature. The consolidation of economic power in the hands of those with access to knowledge and information was essential to the development of American capitalism; and the success of the naval stores industry as it spread along the Atlantic coast was stimulated by entrepreneurs in other states turning to the North Carolinian experts for information. "Several of your subscribers in this neighborhood," a farmer from Alabama told the *Southern Cultivator*, "are desirous of obtaining some information on the manufacture of turpentine—the manner of preparing it for market—the quantity that can be collected in one season by a hand—the expenses attending it—the average prices that it bears in the commercial cities." These subscribers had become interested in naval stores production when they learned "that making turpentine is at present a very lucrative business in North Carolina—far more so than the 'raising' of our present staple commodity." Initially the *Cultivator* suggested that "any one who intends to engage in the business would do well to go to North Carolina and spend a few months in the turpentine district," but in subsequent issues the journal published letters explaining the production process and the business' financial, material, and labor requirements.[76] The *Southern Planter*, agriculturist and author Solon Robinson, the *Farmers' Library*, *Niles' Register*, and that touchstone for the entrepreneurial southerner, *De Bow's Review*, also informed their readers about the profitability of the turpentine business. These journals reprinted each other's technically complex articles, discussing the business in technical detail, breaking down the process into small areas of expertise, and sharing information within an exclusive network of exchange.[77] In the same vein, the organization of agricultural societies, such as that of The Agricultural Society of Bladen in 1823, gave planters a forum, encouraged by legislative act, within which to share current information and knowledge while solidifying their connections with "many of the most respectable planters" in their neck of the woods.[78] As a result, knowledge of the technical aspects of the industry became the intellectual property of a class of literate farmers and planters who devoted themselves to the development of progressive, efficient, and profitable commerce. Trade publications and agricultural societies thus not only informed interested parties of the economic value of the longleaf pine but also affirmed elite perceptions of what constituted proper and improper use, calling attention to the growing "scarcity of this valuable timber" because of "wholesale destruction" and suggesting government protection for forestlands so as to keep the "spoilers" out and allow trees to grow to their full strength and value.[79]

Gavin Wright has pointed out with regard to the southern cotton economy and its development within antebellum American agriculture that a lack of purchasable labor in the North stimulated technological innovation while a system of forced labor in the South drove the expansion of a land-extensive, labor-intensive plantation system, severely retarding technological advancement.[80] In the case of naval stores, however, increased profitability through technological innovation in turpentine processing stimulated the growth of plantations and their slave labor forces in the area of turpentine extraction. Because of a rise in the number and size of market outlets for their produce, farmers saw increased viability in the exploitation of larger portions of their lands and the employment of a larger labor force whose explicit task it was to extract resources from the forest.[81] In communities where forest exploitation for commercial purposes continued to be among the most substantial economic activities, slaves became the primary source of manual forest labor. There, the development of slavery into the rigid gang system that drove the great antebellum cotton boom manifested itself in the increased specialization of menial tasks for black turpentine workers. As turpentine became a hot commodity after 1840, many of the slaves that came to the southeastern corner with their masters found themselves boxing trees, collecting turpentine, and transporting it to a shipping or processing point off the plantation. Their privileges eroding as on-site processing decreased, they increasingly worked exclusively in the extractive phase of the production process. Some slaves still managed to distinguish themselves as skilled coopers, but most worked the orchards full-time: "hands who tend turpentine have no time for any other business," Dugall McMillan of Wilmington informed the readers of the *Southern Cultivator*.[82]

What has perhaps obscured the prevalence of slave labor in the late antebellum naval stores trade is the less rigidly racial division of labor in turpentine operations, both before and after the Civil War.[83] As McMillan pointed out, antebellum turpentine work was the domain of "males, both white and slaves." In one of the most famous—and romantic—descriptions of life on a turpentine plantation, James Avirett depicted a residential blend of black slave cabins and "double log cabins for the families of some twenty or more white men" who were to "guard the orchards from fire."[84] In addition, turpentine work was the kind of labor for which some masters paid their slaves, which suggests that it sometimes fell outside the range of a slave's regular tasks. Yet even if the organization of turpentine plantations did not exactly fit the model of the cotton plantation that was making its way westward and spreading gang-based slave labor along the Mississippi, the "factory in the field" undeniably became the locus of naval stores production after 1840.

As a result, the industry's balance between small and large producers shifted in favor of men like John Mercer, James Grist, and Grist's brother-in-law David L. Russell—planters with sufficient knowledge and capital means to engage in large-scale production. Russell owned 25,000 acres of forestland in Brunswick County as well as 150 slaves, and his estate produced $25,000 in naval stores in 1855.[85] Mercer, born in 1812, moved his family some time after 1843 to the area around Supply, where "with very little capital" he became "a successful businessman and soon acquired large holdings of land, did a large business in naval stores and owned a number of slaves." His plantation, New Supply, included a tide-operated grist mill and a dam, as well as a cooper shop where slaves made barrels for turpentine and tar. During the Civil War he shipped his naval stores down Lockwoods Folly River in small boats. Mercer clearly was a man whose capital, command over labor, transportation skills, and knowledge of technology, tides, and production made him a strong competitor in the naval stores industry until his death in 1863.[86]

James Grist was the son of Allen Grist and the nephew of Richard Grist, two men who had already cut their teeth on coastal commerce before James established large turpentine operations in Brunswick, where he purchased 6,000 acres of pineland, in Columbus, where he purchased a large plantation when he expanded his operations up the river and farther west from the coast, and in Robeson—"for Robeson is the place to make money & this is the year to push hard."[87] He directed his business both on site and from Wilmington, and eventually directed his cousin, Benjamin Grist, to establish another large turpentine plantation in Alabama. His correspondence, especially with his father, reveals the scale of operations: Grist complained that as a result of low water he was "only able to get down 303 bls of spirits turpentine & 500 bls no 1 rosin"—such quantities of high quality naval stores were simply unimaginable for small producers, even under normal transportation conditions.[88]

Letters he wrote and received also attest to the importance of slave labor and the difficulties owners and overseers had keeping their slaves productive: Grist had sent every hand he could possibly spare—"a good lot of boys"—to Robeson so as to take full advantage of high rosin prices, but felt compelled to tell those in charge of operations "how to manage and treat the hands to keep them satisfied."[89] A runaway slave named Abner had come back to the Grist plantation "with a Sithe blade in his hand" and, in

Grist's eyes, deserved the thrashing he got.[90] Allen Grist expressed his disappointment in Bill Perry, whom he considered "one of the beast negrois in Brunswick," who had bitten a merchant's hand.[91] James Grist recounted in exasperation the drowning of Willis, one of two slaves who tried to salvage a few barrels after high water in the river had dislodged them from storage beneath a platform on the river bank: "no good on earth could be done in taking up Spirit Cash—in a small batow rushing down the stream"; the other slave "was a great fool to have undertook it." He was far more pleased with "the boy Gerome," who was "a good Negro . . . and a good cooper," as he was with Sandi, who "behaves & works well" after having been in some "scrape" with a free black.[92]

Despite his own sense of benevolence and righteousness, however, Grist got an earful from a fellow planter, two of whose slaves Grist had rented. One of the slaves, John, had returned to his master in exhausted condition and complained of being "over worked" and "not well fed." The other slave, Albert, also ran away from Grist but had not returned to his master, causing the latter "much anxiety." "Most shocking," he complained to Grist, was "the plan of giving up the management of hands entirely to overseers, who have no feeling"; "confident there was some impropper [*sic*] oppression which drove them to it," he expected Grist to deal with the runaways leniently.[93] One of Grist's overseers, a man named Patterson, complained about a group of chippers who had given him no trouble before but had decided that it was too hot to go to work; it appears he had some sort of sit-down strike on his hands, and he vowed that "all hands will go to work in the morning and I intend to press them from the start."[94] Not only does the large role of overseers in the production process attest to the size of operations, the behavior of the slaves also indicates that they were able to exploit the tension between masters, renters, and overseers that existed on large-scale plantation operations.

James Grist and his fellow planters often shared their concerns, keeping each other appraised of their business fortunes, of market conditions, and of slave troubles. John T. Council, another turpentine farmer, was on at least one occasion paid by Benjamin Grist for services he rendered the Robeson plantation, and wrote James Grist about runaway slaves he had gotten "clear of" after getting one of them out of the Raleigh jail.[95] The Mercers, Councils, and Grists also shared their interest in the turpentine trade's maturation into a complex industry with members of the business community that emerged in Wilmington and the river towns—men who liked to have a hand in more than one aspect of the developing economy. Entrepreneurs like the MacRaes of Wilmington, for example, considered business to be "of vital importance" to their family and were instrumental in the creation of a business system that could handle large quantities of forest produce.[96] Southern equivalents of the boosters who stimulated trade and business in towns across the Midwest, they actively promoted internal improvements in North Carolina: operating from Wilmington, they pushed hard for the incorporation and construction of the Wilmington and Manchester Rail Road, the Carolina Central Railway, and the Cape Fear and Deep River Navigation Company, all of which would open up Wilmington's western hinterland.[97] "This scheme, if properly carried out," John MacRae wrote about the Navigation Company, "would in my opinion be of vast service to the commercial interests of Wilmington. The Central Road too, if properly managed."[98] With these imminent developments in mind, John and Donald MacRae started a general commission business that dealt in, among other things, the naval stores that were about to come to Wilmington in unprecedented quantities.[99] Other MacRae men worked as engineers on canals and railroads and as lumber inspectors on the wharves of Wilmington. They recognized the growing complexity of both technology and markets and actively sought to benefit from it, establishing profitable connections with the proper parties.

The correspondence between planters and other entrepreneurs points to yet another advantage over small producers that larger planters and merchants enjoyed: an informal network of business connections that gave entrepreneurs access not only to extra labor but also to the kind of "social capital" that emerged as a key to entrepreneurial success in capitalist economies.[100] These were the kinds of connections that could help one along within the realm of business, and without which one faced yet another obstacle to entering into the developing industry. A letter to Richard Grist shows the importance of one's standing among like-minded gentlemen. Grist, who had been having the kind of financial troubles that plagued many businessmen in the developing economy (he had been denied further loans "until some favorable change takes place"), had expressed to the letter's author his suspicion that the latter had contributed to Grist's poor standing among creditors. "I aver upon Honour your suspicions are unfounded," responded the suspect; "I have always privately and publickly spoken of you as a Gentlemen and a good Merchant here, knowing the resources of the country, and pursuing the only Branch susceptible of Proffits."[101]

Such connections reached far beyond the southeastern coast and kept planters like Grist appraised of circumstances in the Caribbean, where demand for the products of forest and farm was often a result of environmental circumstances. Droughts and storms could wreak havoc on crops, which could result in increased demand to replace the crop but was more likely to slacken demand because of a lack of income. Successful crops, conversely, often meant great demand for lumber and staves, but lumber was not welcome between June and October, hurricanes being "the greatest drawback on trade." In addition, there was foreign competition to contend with; merchant Matthew O'Neil expected to astonish Richard Grist with the knowledge that "20,000 bus. Corn are daily exported from England, & . . . 10,000 from France." On St. Vincent, Grist also had to compete with imports from British America.[102] Because of the uncertainty of market conditions, ships peddled their wares from island to island, often calling without being able to trade anything, and Grist's business connections would inform him which products he should or should not send to Martinique, Antigua, Jamaica, St. Thomas, St. Kitts, Trinidad, Grenada, or Barbados. James Grist, whose naval stores made their way to such faraway ports as New Orleans and New York City, was in frequent informal contact with merchants such as R. M. Blackwell & Company in New York City and A. H. Van Bokkelen in Wilmington, who invariably informed him that the market was "dull."[103] Wilmington merchants such as R. W. Brown, A. Hewlett, Potter & Parsley, and Aaron Lazarus—who married into the mercantile Mordecai family, built a sawmill in which he employed twenty-five slaves, and led the drive for the Whiteville, Waccamaw & Cape Fear Railroad—also corresponded frequently with Oliver Jones, who kept them informed on the sale of their goods, the state of the market ("verry dull"), and the effects of price fluctuations. At the same time, Jones sent his brother Samuel to Wilmington to report back on the city's business and ultimately to set up a business there.[104] All of this added up to a network of business connections that worked to the advantage of those in it and to the disadvantage of those who were denied access to it because of their suspect social and economic standing.

The services of R. G. Dun & Company, a commercial reporting agency in New York City, were a crucial response to the need for guiding "maps" that would make legible the social "wilderness" that the expansion of capitalism and democracy had created. The agency's correspondents who reported on the credit-worthiness and trustworthiness of businesses in the towns around the Green Swamp relied partly on their own observations and partly on the opinions they solicited from local entrepreneurs. Their reports on participants in the turpentine industry reflect, as a result, the values that underpinned the burgeoning business community's economic and cultural dominance. Large planters, large merchants, and large landholders almost always received favorable credit reports, not just in terms of the capital that could serve as collateral, but also in terms of character and connections. John T. Council, for example, was considered "a capable businessman of large property"; his 3,042 acres of land and assets of $5,866 in 1856 made him "perfectly fit for a northern debt of $25,000."[105] Josiah Moultsby, a Whiteville farmer who also operated a steam mill and produced turpentine, was the kind of reliable entrepreneur who sold $7,000 worth of naval stores per year, owned 6,000 to 8,000 acres of land worth $75,000, and commanded the labor of 75 slaves.[106] Should the son of such a reliable man need to borrow money from his father to get started in business, "this did not hurt his credit much."[107]

Smaller entrepreneurs who had no capital safety net to offer creditors were only considered trustworthy if they were, in the eyes of the correspondent, cut of the same cloth as their industrious, honorable, and well-connected competitors. While the heavily indebted distiller J. G. McDugald of Elizabethtown had "not shown a proper degree of faith towards his N.Y. friends" and was later declared "as dead as a mackrill," Robert & George Fait, also of Elizabethtown, were "industrious honest young men" of little means, but with "good friends and backing."[108] John McKay of Rowland & McKay, a company of "small means," also had "well-off" connections.[109] Traders in Brunswick County, however, were "all small operators, none having a stock worth over $800 to $1000," and "rather a rough set"; the agency "would not advise much credit."[110] Perhaps the best indication of the market's uncertainty in the face of its rapidly expanding frontier is the description of Samuel Sleight of Shallotte, who in 1854 decided to try his hand at commerce: "A young man of small means, just starting a country store. Let your subscribers imagine a young country lad, with little or no education, putting up a store on the edge of the *Green Swamp*, from which the youth himself has just emerged, and they will have some idea of Sleight."[111] While his industry and opportunism may have been admirable assets, Sleight's inexperience and rugged background were obvious liabilities for New York investors.

The agency's identification of the sobriety, reliability, industriousness, promptness of payment, business sense, and honesty of market participants—or their lack thereof—provided highly valued signposts in a developing economy that could be treacherous and hard to navigate. The American frontier, after all, provided charlatans and con men the same kind of mobility and fluidity enjoyed by honest and enterprising men like Calvin Haynes—a farmer, turpentine manufacturer, preacher, and owner of twenty-six slaves in Whiteville who sold his $10,000-a-year business and headed for Texas.[112] Daniel M. Rodgers, whose business was listed as "turpentine," had already been "apprehended in Wilmington for forgery of a check of $20" when R.G. Dun's correspondent took note of him in 1854. Two years later, "worth $00," he was nevertheless "a candidate for sheriff in Robeson." After failing badly, he "ran off to parts unknown."[113] It was also "best to steer clear" of Ardrian van Bockelen, the Wilmington merchant who at one point "stood well socially" and served as mayor of Wilmington but whose "spanking pair of horses" and determination to live "like a game cock" had put his credit "at a very low ebb."[114] Turpentine manufacturer A. C. Dickerson of Bog Swamp, although the owner of $8,000 in land and slaves, was a "strange sort of character," a "rare bird" who told "big lies."[115] And John Carroll of Elizabethtown was the kind of liar without prospects or business sense who, "when he has money . . . you call on him he will say not."[116] The reports of R. G. Dun & Company served to protect the business community from men such as these, entrepreneurs who refused to play by the rules and standards that leading businessmen were defining and implementing as the economy developed and matured.

"In order to insure social union & cast off etiquette"

Although the propertied, moneyed, literate, well-traveled, and well-connected community of planters and merchants had made the naval stores industry largely their own by the time of the Civil War, their dominant position did not simply reflect the polarizing triumph of a selfish capitalist elite's exploitative practices over the environmentally conscious, anti-capitalist, communal practices, and moral economy of small farmers. Rather, they indicate the hegemony of one set of communal practices and values, very much capitalist but still embedded in a larger set of reciprocal human–environmental relations, at the expense of alternative practices and values. White farmers of smaller means, who were not excluded from full participation in this market economy, subscribed to some or all of these values because they felt they could further their own interests, even if they tried to ignore the dictates of the dominant community when they considered doing so profitable, inevitable, or necessary. They pursued not only various levels of subsistence, competency, independence, and leisure, but also integration and participation in the larger Atlantic market, to which Wilmington gave them access, as well as inclusion in the larger political communities—the state and the nation—that were born of the American Revolution. The terms and demands of these larger realms might infringe on local values and activities, but they might enhance them as well, be it by way of access to markets, information, and technology, shared responsibility for internal improvements, or protection from external or internal threats to the community's prosperity and stability. When the state legislature created Columbus County out of Brunswick County because the flooding of the Waccamaw often prevented people living west of the river from making it to the courthouse "without imminent danger," the lack of protest indicates that people considered this kind of government intrusion to be proper and necessary.[117] Similarly, as studies of antebellum yeoman farmers and "poor whites" have shown, the legitimacy of the slave-owning elite's firm grasp on power, the conscription of non-slaveholding farmers into slave patrols, and the tenets of white supremacy were rarely in question.

The predominance of private landownership, especially, as well as the economic and recreational practices it engendered around the Green Swamp, were far less divisive than the patterns of communal land and resource use that were at the core of so many conflicts between states and their engineers on one side and village and peasants on the other. Along North Carolina's southeastern coast, landownership and resource use generally facilitated sustained forest accessibility. As Gavin Wright has shown, most southern farmers, especially those with limited labor power at their disposal, owned more land than they could cultivate with the labor power available to them.[118] This was, in their eyes, not a deficiency—they saw the woodlands they owned as a security possession from which they could win turpentine, timber, pasture, or cropland when necessary. Only when they invested heavily in slaves, whose value exceeded that of forest resources, did it make sense to commit that capital to the exploitation of the forest on such a scale that its returns would exceed the investment. Most farmers, however, did not own any slaves; most slave-owning farmers owned fewer slaves than necessary to help them transform their vast woodland holdings into profitable agricultural fields; and those who owned larger numbers of slaves also tended to own large enough plots of land still to allow them to leave much of it uncultivated for extended periods of time. In addition, land holdings tended to be larger than most farmers could effectively police, despite a rising number of "no trespassing" notices in the newspapers. Although most forestlands were in private hands, they remained to a large degree uncultivated, unfenced, and uncontrolled, leaving vast expanses to serve a more or less communal purpose.

In addition, when the North Carolina legislature in 1825 granted to the newly created Literary Board those wetlands "that could not be utilized for home building without expensive artificial drainage," many of the southeastern corner's swamps, marshes, and bays became public properties whose loose management gave the region's inhabitants further access to wild nature. The eventual drainage and sale of these wetlands—parts of the Green Swamp among them—was to generate revenue for public education. To this end, the General Assembly passed an act calling for the Board of Internal Improvement to survey one or more of the state's large swamps or marshes and determine the practicability, cost, and potential result of drainage. "Instead of being the abodes of reptiles and howling beasts, the receptacles of stagnant waters, which spread disease and death through the country," the engineer chosen for the task concluded in typical fashion, "these now loathsome marshes and dismals may all be converted into fruitful fields." Actual drainage projects did not commence until a decade later when engineer Charles Shaw received sufficient funds to purchase two levels, a theodolite, a compass, chains, rods, and level staves, and to hire a crew of slaves and free blacks. For the wage of one dollar per day, free blacks worked as cooks and carpenters, while slaves, for the same money, wielded axes and chains as Shaw surveyed and mapped the state's wetlands for drainage purposes.[119]

However, when Shaw's successor, civil engineer Walter Gwynn, received an additional $200,000 in state funds and managed to put up for auction, in November of 1841, a number of drained lots on Pungo Canal in Hyde County, it became clear that eastern North Carolinians did not necessarily share the state's opinion that the swamps in the vicinity of their communities had to go. Few people showed up for the auction, even fewer made bids, and none of the offers was satisfactory to the directors of the Literary Board. People were unwilling to pay $6 per acre of drained land only to live amid still undrained wetlands when the swamp's resources were there for the taking in the absence of any real state supervision. Among the few people who showed interest in purchasing swampland was Whiteville farmer Fourney George, who had surveys conducted for four tracts comprising 5,000 acres in Columbus County and concluded that one of these tracts—2,000 acres of flat, undrainable "bay" country—was worth only five cents per acre. "The only growth that gives value to it is Cypress," he wrote Governor Thomas Bragg in 1855, but the stands were "not sufficient . . . to make any considerable investment or outlay . . . So far the shingle business in this swamp has been found not to pay. The cost and trouble of getting them to market are too great."[120] People were more comfortable occasionally extracting resources from the swamps around them than they were owning one. Accordingly, the majority of the few sales the state managed to make were of better-timbered swamplands.[121]

The longleaf pine forest and its many swamps thus remained extensive enough for the maintenance of traditional ways of life, subversive sustenance, and recreational hunting. Nature offered many an avenue of respite and, in the case of runaway slaves, escape and refuge. It was the kind of environment where the Cumberland Fox Club could "challenge their brother Sportsmen" from neighboring counties to "a three days' Hunt," or where a local militia company could declare "a war . . . against the squirrel tribe" and proceed to kill 1,500.[122] Little wonder that traveler Basil Hall observed that the free population of the region "prefer hunting."[123] On the occasion of snow and frost, people went on sleigh rides and seized "the opportunity for skaiting."[124] Catherine Clitherall, wife of a planter on the Cape Fear, described how her family joined other planters at a vacation resort in the piney woods where they played at being common people: "In order to insure social union & cast off Etiquette or any approach to luxury," the families agreed "that the Houses erected there should be 'log houses.'" (Never mind that slaves brought fruits, butter, and poultry up from nearby plantations; that management delivered lightwood to the cabin door; or that "a male serv[ant] walk'd in front, with a blazing torch necessary to guide one thro' the stumps and roots" when the fourteen families took to visiting one another at night.[125])

Although having a vacation home in the woods and the time to escape to it when the summer heat became too oppressive were luxuries only the wealthy could afford, planters and farmers nevertheless shared in their daily lives on the forest's edge a proximity to the nature that provided them their daily bread. The diary of Elizabeth Ellis Robeson, a planter's widow of Tar Heel, Bladen County, shows how rural people interacted with nature on a daily basis. Faithfully noting, like so many farmers who kept diaries, the occurrence of rain and floods, heat and droughts, and snow and frosts, as well as how they affected her crops, she was aware of her dependence on what she considered nature's providential blessings. A pious woman, she discerned the hand of God in destructive floods and the many sudden deaths of loved ones and acquaintances, often reminding herself not to question divine motivation. Yet she endowed annually recurring activities, such as the hauling and shucking of corn or the killing of hogs, with little ritual meaning; she took for granted the wide availability of nature's resources for recreational or economic purposes when she and her children went fishing, "banked a tar kiln," or "started a raft to town."[126] Catherine Eliza Clitherall too enjoyed the kind of abundance so many farmers strove for and, to varying degrees, achieved by "raising sufficient Poultry" and "having a plentiful dairy, fine garden . . . superfluous rice for fine Turkeys," and a nearby river "abounding with fish and wild Ducks." Going "to the Oyster beds, upon which we sat & roasted our oysters," was another widely shared pleasure when the coastal waters still abounded with easily obtained food, and access was not yet strictly policed.[127]

Neighbors joined one another in church, where they gave communal expression to their private fear and awe at the powers of nature that not only provided for them but could turn on them as well. In 1826 citizens of Wilmington, believing that only divine intervention could appease the force of nature, observed "a day of general Fasting, Humiliation and Prayer, in consequence of the severe drought under which they were suffering." The *Carolina Observer* considered it "a providential circumstance" when a particularly violent storm claimed no human lives.[128] Such sentiments and activities were born of the feelings of awe, fear, humility, and uncertainty that were common among nineteenth-century Americans of all classes. When Catherine Clitherall sailed from England to North Carolina in November of 1801, she experienced "a storm so terrific it can never be erased from my memory," and thanked "The Providence of our Heavenly Father" for blessing those on board with the presence of a passenger who took charge when the ship's captain proved incompetent and mentally unstable. When the storm subsided, she went on deck "to take advantage of the opportunity . . . of viewing one of the most sublime spectacles in nature, that of the troubled ocean gradually recovering from its turmoil, and regaining its general tranquility." At first the sight was "awful," she confessed, but after some time she and a "piously educated" fellow passenger were "able to appreciate the work of God, so wonderfully manifested to us."[129] Some years later, when a violent storm raged around the plantation while her husband was sailing for Charleston, she spent the night "in prayer," and attained "a consolatory feeling that God was powerful . . . & could still the raging storm."[130] And such sentiments were not the exclusive domain of supposedly frail Southern belles: on 19 October 1848, Elizabeth Robeson recorded in her diary that the night before "the Aurora Borealis was to be seen. It looked *grand and awful, a great many alarmed at it.*"[131] To a large degree, the people of the Inner Coastal Plain Region shared not just an interest in the exploitation and improvement of the natural world that could frighten them so, but also the faith-based worldview that provoked these fears and directed their interactions with nature.

* * *

Although economic development took a heavy toll on the forest and wetland environment around the Green Swamp, the majority of its inhabitants did little to impede the rise of a capitalist business community with such a large hand in the direction of resource use and economic development: slaves because they could not, at least not openly; Indians because they were too small in number to stake a serious claim; free white people, male and female, because they welcomed for the most part the benefits that came with what they understood to be a proper and necessary attempt to turn the wild wastes of the Green Swamp region into productive farms by way of a reasonably fair system of opportunity and competition; and all of them because the environmental impact of economic development, mitigated by modern landholding patterns and traditional patterns of forest resource use, was limited enough to leave enough forestland and swampland "untouched" for the landscape to retain its economic potential, its recreational attractiveness, and its inhospitable, aesthetic, and spiritually imposing character. Still, developments in the naval stores trade—driven by waterways, railroads, and cities—enabled a growing number of settlers to move further into the woods and commence shipping out their natural resources. They set the southeastern corner's mixed economy on a capitalist course, steered by a business community of elite planters and merchants who used their command over investment capital, slave labor, processing and transportation technology, market information, social capital, political power, and knowledge of the larger world to dictate the terms and police the boundaries of proper and legitimate resource use in a legible landscape. Soon, the capitalist developments that drove the naval stores business would come to even greater fruition in the lumber industry, which solidified the socioeconomic dominance of the business community and further circumscribed alternatives to the dictates of legibility and order that accompanied and facilitated capitalist development.

3

"Enterprise vs. The 'Grab Game'"

In 1868 Lemuel Bannister, a businessman from Cleveland, Ohio, traveled to Brunswick County to take his position as the president of the Green Swamp Company, a lumber company operating from the town of Bolton. Soon after his arrival, he and "several gentlemen" ventured into the Green Swamp, across Lake Waccamaw, and down the Waccamaw River, on a reconnaissance of the company's holdings. The wilderness and the people he encountered confronted this Northerner with a South that was, as he put it in a letter to his fiancée, "unfinished." Beating a path through the swamp's dense woods, the party came across "the wildest and . . . most magnificent scenery." They also came upon "an aged woman belonging to that famous tribe . . . known as the 'Poor Whites'" who, despite the swamp's impending development, refused to leave her solitary home on Crusoe Island. This was the kind of wilderness Bannister had come south to cultivate: although a "dark," "forlorn," and "naked hermitage remote from all the pleasures of the world," one that aroused in him a spirit of adventure, the Green Swamp also appeared to Bannister "a mine of wealth" with its "valuable and inexhaustible timber." And because Southerners were, as he wrote his fiancée, "not as devoted to gain, or as shrewd in business affairs as they might be," he would have to take charge if anyone was to profit from this wealth. In order to harvest the riches, however, Bannister found himself having to deal with those backwoods dwellers as well as with newly freed black workers. This he found rather frustrating, and harder to get used to than the North Carolina heat, which resembled "the regions of his Satanic majesty."[1]

Like so many enterprising individuals in the years after the Civil War, Bannister saw economic opportunity all around him but struggled to build a business as he ran into the obstacles the wilderness and the communities around it presented. Already for more than a decade before his arrival, the founders of the Green Swamp Company found North Carolina's laws of incorporation much more accommodating than its timberlands.[2] By

1903, thirty-five years after he had first marveled at the wild beauty and economic potential of the Green Swamp, Bannister's resolve to harvest its riches had only hardened in the face of the company's inability or unwillingness to do so. He co-authored a report whose sole focus was the value of lands surrounding Lake Waccamaw to lumbering and agricultural interests. Waterways, he wrote, "can be made available at slight cost, to raft an immense quantity of Cypress, Yellow Pine, Juniper and Gum Timber to Wilmington . . . and there manufactured and shipped to the various markets on the Atlantic sea-board as well as to the foreign ports." The land, once cleared of trees and properly drained, would, in his estimation, make good farmland for the cultivation of agricultural crops. Gone was the romanticism of his earlier wilderness explorations—"value," "cost," "market," and "profits" had become keywords to a man who no longer reveled in the beauty of the wild swamp but instead anticipated its transformation, through lumbering and drainage, into profitable farmland.[3] Yet the lands that once belonged to the Green Swamp Company had long passed into the hands of other lumber companies. The company, like many others, had probably found traditional methods of logging and rafting insufficient to generate a profit in the dense swamp and lacked the capital to mount large-scale logging machinery. Not until the Maryland-based Waccamaw Lumber Company erected a large-capacity sawmill in Bolton and employed the necessary labor and machinery to build several railroads into the swamp did logging there prove profitable.

Building on the groundbreaking work of C. Vann Woodward, historians of the South's postwar economy have forged countless stories like Bannister's into a "New South" narrative of spasmodic capitalist development circumscribed by socioeconomic attitudes and social relations harking back to the antebellum era.[4] Like Bannister, many Northerners came south in the years after the Civil War to make their fortune in an extractive industry that helped transform an underdeveloped rural region into a more developed agricultural/industrial hybrid. Benefiting from the wide availability of cheap public lands to amass huge holdings of timber, coal, and ore lands, they then utilized the region's cheap, often seasonal labor to turn them into the resource base for the industrial sector that emerged amid the plantations and small farms of a largely agricultural economy. They had to operate in a labor market whose sudden transition to black wage labor forced white entrepreneurs to negotiate with black laborers, a situation that developed in a way that neither party much appreciated.

The twin pillars of race and slavery loomed large over southern economic development, as seen in the rapid rise of sharecropping and convict labor, in the emergence of company towns whose organization historians have likened to both the paternalistic and the oppressive character of the slave plantation's power structure, and in the scarcity of biracial labor unions and other organizations or movements whose advocacy of economic equality and integration threatened white supremacy. Also, fueled by northern capital and business know-how, companies like the Green Swamp Company and their black and white laborers laid to waste much of the region's natural environment when they harvested and processed its natural resources with little regard for environmental preservation, future supplies, or the well-being of workers. Characterizations of industrial development as dehumanizing and environmentally destructive reverberate across the historiography of postwar southern industrialization.[5]

Historians of southern lumbering have generally cast forest industrialization as the emergence of a powerful economic sector through deliberate, destructive, profit-driven clear-cutting that constituted a clear break with the past. Geographer Michael Williams has characterized the forty years between 1880 and 1920, during which lumber companies rapidly leveled millions of acres of forest, as a "lumberman's assault on the southern forest." Like Henry Clepper, who spoke of the "plunder of the pineries," Williams painted a picture of violent invasion and abuse to condemn the industry's methods of exploitation.[6] Indeed, most historians who write about southern lumbering use such terms as "wasteful," "loss," "desolation," "scarred hillsides" and "depletion" to describe its effects on the environment, as well as "grueling" labor and the "horrible suffering of mules and oxen" to characterize its toll on humans and animals.[7] Depicting the industry as an all-consuming capitalist force moving across the American landscape, they evoke repeatedly the "cut-out-and-get-out" mentality of mill owners to emphasize the industry's transient nature, echoing the kind of criticism Assistant Forester Earle Clapp of the U.S. Forest Service leveled at the lumber industry as early as 1920: "Each successive chapter in the history of the lumber industry in the United States has been a story of depletion and migration," he wrote; the South, "following the course of other regions," was just another stop on the path of destruction.[8] The lumbermen of the New South, it follows, were the ruthless agents of capitalism who may have believed that their company towns were models of paternalistic management but who generally trampled all over pre-capitalist yeoman farmers, powerless lumber workers, state and federal conservation agents, and nature itself.

A proliferation of sawmills does not, however, a monolithic "lumberman's assault" make. Although the booming lumber industry as a whole undeniably wreaked tremendous environmental havoc in the southern pine belt, we must be careful not to ascribe to the industry's individual components the same homogeneity of size, structural organization, and socioeconomic and environmental impact that informed *The Visible Hand*, Alfred Chandler's groundbreaking thesis that canonized the "managerial revolution" as a model for business history.[9] As Philip Scranton shows, the rapid, widespread acceptance of Chandler's argument regarding a small number of really large companies created a "silence" that enveloped "the other 80 or 90 percent of the nation's industrial capacity." Touring "infrequently visited sections of the American industrial terrain," Scranton illuminated the important role played in the process of American industrialization by the countless companies engaged in specialty production of goods that proved crucial for the expansion of industrial productivity and individual consumption. Emphasizing the importance of "the ideas and actions of 'ordinary' proprietors and workers" to the economic stimulus specialty production provided, he also called into question Chandler's reliance on "factors," processes, and structures, rather than human agents.[10] Further challenging Chandler's paradigm, AnnaLee Saxenian has posited as important alternatives to the omnipotence of the modern corporation the "regional network-based industrial systems" that these ordinary proprietors and workers forged with other specialized companies.[11] Although the industrial development and economic impact of large corporations are extremely significant, Chandler's detractors contend, they cannot serve as the exclusive models of industrial production and organization in the United States.

Although the rise of large lumber companies mirrored in some ways the emergence of the modern corporation, the extractive nature of the lumber industry prohibited the development of individual companies to the size, complexity, and permanence of Chandler's industrial giants. Timber remained too bulky and too widely available for any one company or conglomerate of companies to become powerful enough to monopolize logging, transportation, sawing, and marketing. Large corporations, therefore, continued to share the forest with small companies. Moreover, lumber production, although increasingly mechanized, remained too basic a process to require the kind of structural reorganization and salaried managerial class that reshaped the far more complex industries that produced steel and chemicals. Inevitable resource depletion also prevented the permanent establishment and full integration into local economies of companies whose basic function was resource extraction. Whereas a corporation like U.S. Steel could afford to bring in anthracite coal over growing distances because of the much higher value of its end product, it made little sense for lumber companies to sustain the capital investment of a sawmill once nearby timber supplies had been exhausted. And the technological revolution that drove the rise of the modern corporation only exacerbated the transient nature of the lumber industry: the more advanced logging and sawing technology became, the faster companies depleted their resources and moved on. As a result, shifting clusters of large and small companies, rather than fixed giant corporations, characterized the early development of the lumber industry.

The size, organization, and environmental impact of such lumber giants as Weyerhauser and Kirby, then, although they approximated those of other "captains of industry," cannot serve as the model for the development of networks of small sawmills, auxiliary industries, and local merchants in an area like the Green Swamp region. In a 1924 report, the North Carolina Department of Conservation and Development classified the state as a "small mill state" and attached "very little significance . . . to the number of mills cutting out in a stated period."[12] While a number of outside entrepreneurs did bring to the Green Swamp the capital, technology, and labor typical of America's rising big business, effecting such a drastic expansion of timberland ownership and extraction that the forest under their control underwent an equally drastic transformation, these local operators of small mills did not participate in what they called a "grab game." Instead, they championed "enterprise" as "the only sure means of building up a town."[13] They followed local, established patterns of development that facilitated the emergence of a business community that may have shared many of the values outside entrepreneurs sought to impose on the landscape, but that also mitigated the social and environmental impact of economic development through landowning patterns, social relations, and communal interests. They tapped into the infrastructure of forest use that had evolved before the war, used local labor, and continued the long-standing tradition of general merchants/planters who not only were involved in several business ventures at once but also took part in the recreational activities that characterized rural life on the forest's edge. Within the communities that the lumber industry transformed, these local companies, with their diversified operations, managed to attain a degree of permanence and coexistence with local environmental practices.

"Mill property is not a paying business here"

In 1884 the federal government published Charles Sargent's extensive survey of the nation's forests and woodlands; its section on North Carolina contained a map that clearly showed the limited headway the lumber industry had made in harvesting the state's timber resources, while shedding light on the obstacles to profitable exploitation lumbermen still faced. The vast bulk of cutover land was located along the state's main waterways, which served as primary transportation routes. "The merchantable pine," Sargent wrote about the southeastern corner, "has been removed from the banks of the Cape Fear and other rivers flowing through the southern part of the state, and although these streams still yield annually a large number of logs, they are only procured at a constantly increasing distance from their banks and with a consequent increasing cost for transport." Apart from a long strip of cutover land along the Wilmington and Weldon Railroad, the forestlands located at a distance from the main waterways remained rich in timber because transporting large trees over long distances required too much effort and added costs lumbermen could not hope to recover in a domestic market circumscribed by widespread individual access to woodlands. Lumbermen, understandably, were anxious to find ways to simplify and make profitable the cutting of an estimated 1,581,000,000 board feet of longleaf pine still standing in Bladen, Brunswick, Columbus, and Robeson counties.[14]

This environmental and economic state of affairs was the result of lumber's slow maturation as a profitable commodity. Much more than in the case of turpentine, the creation of a lumber *industry* entailed the transformation into a viable commercial endeavor of a primarily domestic activity that lacked local and internal stimuli for commercialization. Although settlers, especially large landowners along the main waterways, had rafted timber to the Wilmington market since the earliest days of white settlement, most people in the Lower Cape Fear region prior to 1880 cut the timber on their land for domestic and farming purposes—it was largely for this reason that planters equipped their plantations with water-powered sawmills and provided sawing services to others. Many farmers cut trees and set the fallen trunks on fire; some girdled trees a year prior to the cultivation of their land and used fire to level the dying trees; others simply put the torch to their standing timber. As the South's agricultural literature indicates, the need to clear land was much more immediate than the desire to sell its stumpage, especially since the large number of settlers who owned wooded land inhibited the development of a domestic lumber market. Hence, farmers concerned themselves with questions regarding the best time to cut timber so as to ensure its preservation and endurance in domestic applications. They also discussed the feasibility of small portable steam sawmills for the efficient clearing of their own farmlands, the usefulness for fertilization of burying logs in the ground, and the possible merits of lumber wagons and mechanical stump pullers.

Because clearing even an acre with axe or saw was a Herculean feat, especially without the use of hired or slave labor, fire was an especially effective way to clear land in the longleaf pine forest. Janet Shaw, a visitor during the revolutionary era, noted how "a fine wood was set on fire"; shocked at the sight of 20,000 trees gone up in smoke, she "was informed of the reason and necessity of it."[15] To the participants in the forest economy, facing a timber supply that seemed endless and widely distributed while the labor supply was clearly limited, such destruction of potentially valuable timber was efficient and necessary. To farmers who lacked the means to produce and transport potentially rewarding quantities of timber to a nearby shipping point or market, the energy they expended felling trees, preparing them for transport, hauling them to a river, rafting them downstream, and traveling back up seemed much the bigger waste. Given the means of cultivation available to most farmers, the expanse of the forest and the size of the tall pine trees impeded commercial forest exploitation.

As a result, even though occasionally cutting and rafting a few trees was an option open to many, only entrepreneurs who, as in the turpentine trade, had sufficient access to timberlands, labor power, market information, transportation facilities, and sawmills could take advantage of the conditions that between 1800 and 1880 facilitated the steady growth of Wilmington's lumber market. As an increasingly thriving port city, Wilmington was home to a growing, timber-consuming shipping industry. The Wilmington harbor every week saw the coming and going of vessels on their way to and from the Caribbean, Europe, the Deep South, and northern Atlantic ports. Moreover, a looming timber shortage in the West Indies turned all the ports of the southern Atlantic seaboard into suppliers to the Caribbean. After a great fire swept through Port-au-Prince on 16 January 1823, for example, coastal lumbermen were urged to take advantage of the Haitian government's decision not to impose duties on badly needed lumber and other building materials.[16] The resurgence of the naval stores industry also created a strong demand for barrels and staves; urban growth translated into a surge in construction projects and a demand for lumber and firewood on the part of a growing number of urbanites who did not own timberlands.[17] The railroad network that emanated from Wilmington after the 1840s consumed even larger amounts of timber for crossties and fuel.

Entrepreneurs looking to profit from these developments needed to keep up a steady flow of lumber to the Wilmington docks, preferably those kinds required for specific applications, such as construction and cooperage. They could raft timber down the Cape Fear and have it sawed at one of five steam sawmills in Wilmington or one of two water-powered mills fifteen miles further downstream, but rough timber did not command high prices and the sawing came at a cost.[18] Much more profitable was sawing and planing at one's own mill and rafting boards and flooring to Wilmington, where it was sure to command much higher prices. Moreover, pre-sawed lumber could be rafted in larger quantities, so profit margins increased drastically for producers who did their own sawing and rafting, especially those who knew which types of lumber were in demand.

Consequently, the lumber trade underwent changes similar to those that consolidated the turpentine trade into the hands of an entrepreneurial planter/merchant class. James Purdie of Bladen County, a planter who sawed the timber on his plantation into lumber for the Wilmington market; Wilmington's Aaron Lazarus, a dealer in naval stores and lumber who used his slaves to handle the trade's bulky materials; and the Turlington brothers, William and A. J., who operated both in the city of Wilmington and in Columbus County to coordinate the dealings between the city and its hinterland—these were among the entrepreneurs who created and operated a business network of lumber production, inspection, marketing, and trade that left small producers increasingly limited access to the main channels of regional commerce.[19] Alexander Elliot, a Fayetteville planter, exemplified this class of men and their dominant position in the trade. In 1819 he owned nearly 2,740 acres of land in close proximity to creeks and rivers as well as a steam sawmill.[20] He also owned fourteen slaves, several of whom did the logging on his plantation at wages of fifty cents "for labour done nights and Sundays."[21] Using so-called timberwheels, they hauled logs to the sawmill, where they sawed boards and flooring.[22] Then, using either clamps or a technique called "staking," they built a raft of the lumber and rafted it down the Cape Fear: between 1823 and 1861, Elliot shipped at least thirty-five large loads of lumber to Wilmington. Initially, once the lumber arrived in Wilmington, Elliot sold it or had it sold directly to general merchants from whom he bought groceries and merchandise. But at some point the planter started to call on the services of some of the commissioned agents who had begun to act as specializing intermediaries in the lumber business.[23] Once the lumber arrived and Elliot's agents had located a buyer, an inspector appraised the lumber so as to assure a fair price. These inspectors graded and classified the lumber, then had the lumber re-rafted to prepare it for transportation by the buyer.[24] After the agents closed the deal, they subtracted wages and commission and sent Elliot a check for the net profit, often including a report on market conditions and suggestions regarding what kind of product he should ship next.

Although agents almost unfailingly reported that the market was "dull," "extremely dull," or "dull dull," planters who kept in frequent touch with agents and merchants were much more in tune with price fluctuations, shifts in market demand, and the subtleties of the grading system. For example, Elliot received the following request: "If delivered during the month of May $5½ pr. m. ft. can be had for the following: 25 m. ft. 2½ by 8 in 18 ft. long; 25 m. ft. 2½ by 9 in 20 ft. long; 25 m. ft. 2½ by 10 in 22 ft. long . . . to show all heart on one side and to be square edged."[25] Not only were small farmers unlikely to receive such requests, they were also unlikely to be able to comply. Men like Elliot had the knowledge and the resources to understand and respond to such demands, a position they only stood to fortify when they looked for wives among "young widows and old girls owning large plantations and sawmills."[26] Moreover, profitable sawmills were in the hands of similarly propertied and well-respected gentlemen in Wilmington: Stevenson and Hooper, the former being worth $20,000 in 1854; Potter and Kidder, who owned at least $50,000 in real estate, slaves, and bonds in 1848, and whose "first rate" steam mill by 1853 made as much as $30,000.[27]

Despite the consolidation of the means of resource exploitation in the hands of men like Elliot, the lumber industry did not take immediate advantage of the technological advances that facilitated the turpentine industry's expansion in the decades before the Civil War. Although the railroad would ultimately prove the greatest engine of change for the industry, the network that emanated from Wilmington initially served to boost the manufacture and transportation of turpentine; despite the availability of railroad facilities since the early 1840s, lumbermen depended on waterways longer than did turpentine men. Several decades into the postwar era, steam continued to power the vast majority of sawmills, while the technology to load large quantities of lumber onto railroad cars remained too expensive to obtain or too cumbersome to operate in woods and swamp. Most of the mills around the Green Swamp, therefore, remained small affairs prior to the 1880s, with maximum capitalization of a few thousand dollars. Unless they were part of a mixed business, as was the case with Dewmill & Foster's successful dry goods, distillery, and sawmill business during Reconstruction, they usually folded quickly. Hiram Lee of Thompson, Robeson County, had taxable property of only $2,480 when he sold his sawmill and left for South Carolina in 1879. J. Wessel's Lumberton sawmill was capitalized at $8,000 in 1874, but failed to make money, and by 1878 was heavily mortgaged. The failure of K. Phillips to turn the business around after he invested a portion of his property in Wessel's mill led R. G. Dun's correspondent to remark that "mill property is not a paying business here."[28]

Only shingle producers managed to make some inroads in the swamps, and the size of their individual products enabled them to seize as well the opportunities the railroad offered. J. S. Springer's shingle production in the Green Swamp, forty miles from Wilmington, was capitalized at only $5,000 in 1872, but considered "very good."[29] Charles Oscar Beers came to Bolton from New York in 1868 with ambitious plans to build a railroad and workers' shanties in the Green Swamp in order to manufacture lumber. "Regarded as a good energetic businessman and much respected" but without property of his own, Beers became the silent partner of the "perfectly reliable" H. B. Short of Flemington, whose estimated worth increased from $15,000 to $50,000 in less than two years as the two men turned their combined talents to the development of real estate and the manufacturing of shingles and lumber. Like other entrepreneurs with limited technological means, however, Beers found that the swamp's ruggedness allowed only shingle production on a large scale. While his lumber output remained sporadic and limited until he constructed a sawmill near

Hallsboro in 1881, by 1869 he had managed to fill in the mouth of a canal near the wharves he had built—thereby lowering the water level near his place of production to accommodate logging—and was already shipping loads of 28,000 or 38,000 shingles by rail.[30]

"Messing about the mill"

Given the slow and uneven development of the lumber industry despite the optimism and opportunities of the postwar years, the Civil War, that great divide between Old South and New, was not the watershed event that pushed the Green Swamp region's lumber industry full-blast into the racially charged maelstrom of land, labor, and northern capital that enveloped much of the former Confederacy. For about a decade and a half after 1865, the men who had dominated Wilmington's business life before the war only sparingly expanded into the city's hinterland an industry still very much hampered by environmental and economic conditions. The North Carolina Land Agency, a dealer in southeastern lands for agricultural and industrial purposes, alerted prospective buyers that the Civil War had created "almost everywhere in our country . . . a great demand for lumber for building purposes." This situation, the agency claimed, proved very profitable for citizens "fortunate enough to be able to get possession of an old sawmill, on or near the line of our railroads and rivers."[31] However, the agency's activities were more indicative of the rabid land speculation among northern interests that swept the immediate postwar South than of actual economic development.[32] The road to the postwar lumber boom was paved with the failure of companies like the Clarendon Lumber Company of Columbus County and the Southern Saw Mill and Lumber Company of Robeson County. The former purchased for $775 the rights to the "standing and growing *and* dead and blown down" timber on 501 acres of private land, only to sell them to the Whiteville Lumber Company after six months of failure to cut any of the 275,000 feet it expected to harvest.[33] The latter, a large company whose assets included an electricity plant that enabled round-the-clock operation, went into receivership less than a year after it assumed operations, having hardly made a dent in timber holdings that by the time of a court-ordered commissioners' sale still amounted to at least eight million board feet.[34]

The early appearance of a northern entrepreneur like Lemuel Bannister was rather anomalous to postwar conditions in the southeastern corner, even if both his desire for large-scale forest exploitation and his failure to achieve it were typical of the times. In an area where no public lands were available for sale by the federal government, most commercial development was the domain of local entrepreneurs. More representative of the early New South were the various enterprises of Bladen County local John Baldwin, who with some success sought to use his home region's abundant woodlands to attain a level of economic comfort in a slowly developing economy hampered by infrastructural constraints. The fact that Baldwin could on one day casually witness the hanging of a black man in Robeson County and attend an African-American religious meeting the next speaks volumes about the explosive blend of newness and familiarity that was the postwar South. But the trajectory of Baldwin's working life is more illustrative of the paucity of internal stimuli for economic growth that had the early postwar South buzzing with commercial activity without palpable, large-scale economic development.

Baldwin, a white man born in Bladen County on 9 October 1845, by his own account grew up "a poor boy" of "poor parents." His parents were farmers, but Baldwin felt a stronger inclination toward mechanical labor. As a white man with some technical skills, Baldwin enjoyed quite a bit of economic mobility, even if it was often only lateral. On 1 January 1867, he started work for the Bladen Land Company in Bladen Station, forty-five miles above Wilmington on what was then called the Wilmington, Charlotte and Rutherford Railroad. "They Runn too Saw Mills Manufacturing Lumber from pitch and yelow Pine," Baldwin described the activities of the company, which owned only a limited amount of timberland around town. He worked as a sawyer and a mechanic until 13 February 1867, after which date he worked off and on as a mechanic and as a blacksmith, and engaged in farming and repair work until on 18 March 1867, he once again joined the Bladen Land Company. After that second spell with the company he changed jobs several times, moved from Bladen to Wilmington and thence to Robeson County, and eventually returned to Bladen to set up shop as an independent mechanic. "Comensed work for myself at Bladen Station," he noted in his diary on 8 September 1868; the next day, having bought 500 feet of lumber from his former employer, he "comensed building a shop" for himself.[35]

Much of Baldwin's mobility stemmed from the persistence of antebellum labor conditions and from the availability of woodlands, access to which was at best loosely circumscribed: these allowed Baldwin to venture out on his own into the wooded rural landscape for work and for play. To be sure, during his second spell with the Bladen Land Company Baldwin made some diary entries that seem to attest to the growing monotony of industrial labor, as well as to the "dumbing down" and "de-skilling" of the workforce that were among its notoriously dehumanizing aspects:

Monday 5 Sawing
Tuesday 6 Sawing
Wednesday 7 Sawing
Thursday 8 Sawing
Friday 9 Sawing
Saturday 10 Sawing
Sunday 11 Went Visiting

Monday 12 Sawing
Tuesday 13 Sawing
Wednesday 14 Sawing
Thursday 15 Sawing
Friday 16 Sawing
Saturday 17 Sawing [. . .]
Sunday 18 Went Visiting

But work in the sawmills of the New South did not differ much from earlier mill labor: long periods of idleness, often due to a shortage of sawing timber because mills owned little stumpage and lacked the means to obtain more logs, regularly interrupted spells of heavy sawing.[36] Baldwin passed these slack periods "messing about the mill" and frequently spent half-days in the woods to fix a sawmill or go fishing. When he set up a little sawmill as part of his own business, he often "worked . . . in the swamp getting seedar timbers"; although he did not own timberland, Baldwin found plenty of logs in the woods around town. In addition, Baldwin went into the forest for diversion, religion, and reflection. "Went out in the litwood snags woods hog hunting—Returned and went to Preaching at night," he wrote in February of 1871. Other times he went fishing, "gunning," or "riding on hors [*sic*] back viewing the water."[37] In this he was no different than George Nelson Hill, a Brunswick County farmer who combined the toil of cultivating the soil with fishing, killing snakes, and hunting hogs, deer, and turkey—enjoyable activities that at the same time reinforced the value of nature as provider and of man as its steward. Hill, too, was drawn to scenes of nature's grandeur: he specifically traveled to the Green Swamp to see up close the site of a great fire he had witnessed from his farm while it was raging.[38]

"Waccamaw lumber company owned everything in sight"

While the ecological impact of more than 200 years of such non-native settlement and resource extraction was far from inconsequential, in reach and intensity it fell far short of the changes wrought by the modern lumber industry that emerged during the 1880s. In terms of the environmental change stemming from the industrial developments it encompassed, that decade constituted far more of a watershed than the Civil War for the lumber industry and the forests it took on. When the once vast white pine stands around the Great Lakes showed serious signs of depletion and long-held notions of yellow pine's comparative inferiority proved erroneous, the lumbermen of the Midwest turned their covetous gaze to the southern pine belt. Armed with the capital to build numerous logging railroads, apply modern lumbering technologies, and hire large labor forces from beyond the forests' neighboring communities, they rapidly moved and expanded operations in order to harvest, process, and export previously inaccessible timber resources. Able to overcome the natural obstacles to economic growth that had long frustrated local lumbermen, they drove their extractive machinery deeper into the woods than had previously been possible or viable in order to meet industrializing America's exploding demand for railway ties, telegraph and light posts, shingles, furniture wood, and building materials. After 1880 the forests of the eastern seaboard helped feed a southern lumber boom that in North Carolina alone left 10,200,000 acres of cutover pine land and drastically altered the forest environment. "North Carolina, which once had large areas of longleaf pine and was famous as the 'Longleaf Pine State,'" reported the U.S. Forest Service in 1920, "can boast now of hardly 50,000 acres of second-growth longleaf pine widely scattered in small areas."[39] And in addition to longleaf pine, lumber companies also removed enormous quantities of yellow poplar, black walnut, juniper, cypress, and loblolly pine from all corners of the lower Cape Fear area.[40]

The key to the lumber industry's development into a big business was mobility of capital and labor, which ended the dependence on the centrally located, water-powered mills that had long limited the radius of lumbering activity. Although liberal federal land policy in some southern states had allowed the size of individual lumber holdings to approach one million acres, the large cost of transporting bulky logs over long distances had continued to limit the size of actual logging operations. When the development of mechanical loaders made possible the relatively quick loading of large quantities of timber onto trains, the railroad quickly became the single most important driving force behind the burgeoning lumber industry: the southern lumber boom was made possible by a fourfold growth

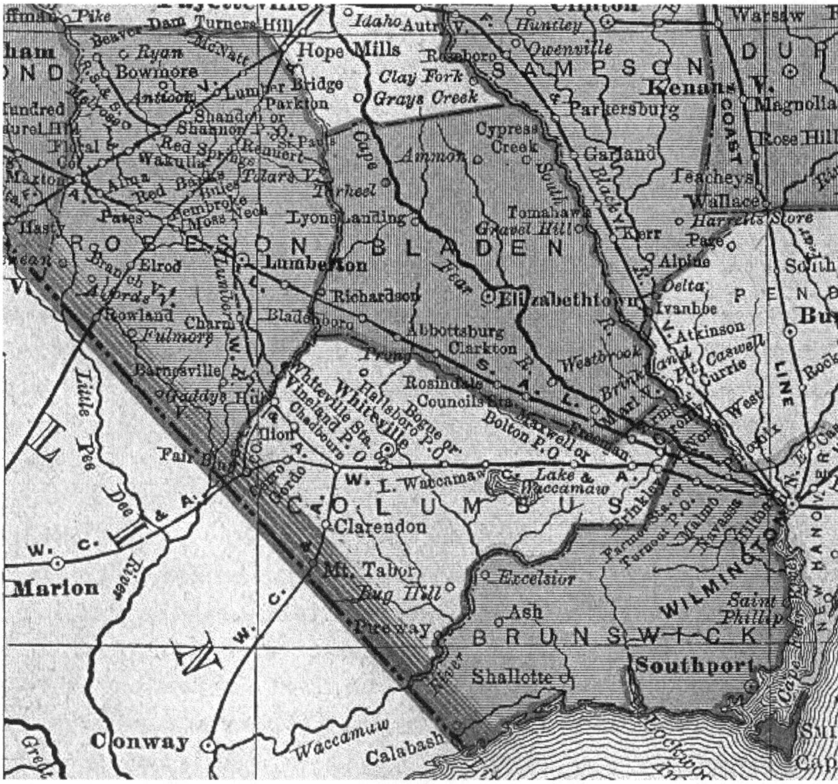

Figure 4. Railroad map of southeastern North Carolina, 1896. The map shows the network of railroads cutting through the area and the various towns that had sprung up around the Green Swamp. (Courtesy of http://fermi. jhuapl. edu/ states/1895/nc_1895. jpg.)

in mainline railroad mileage, from 9,183 miles in 1865 to 38,893 miles in 1910.[41] These railroads, and the many moveable logging lines that branched out into the surrounding forests and swamps, opened up to commercial exploitation millions of acres of woodland that had previously been too far removed from navigable waterways to allow profitable logging with mules and oxen. They also stimulated the growth of small railroad towns and depots which served as agents of economic development like the river towns of the early nineteenth century. In a span of two decades, the construction of a number of similar branch railroads in the southeastern corner stimulated the growth of a network of mills, towns, and tracks that divided a vast forest into smaller segments for more efficient exploitation: Between 1880 and 1884 alone the number of saw- and shinglemills listed in lumber company directories jumped from seven to fifty-two.

The success of the railroad in stimulating logging operations in the longleaf pine belt's swamps, barrens, and savannas received a boost from the development of additional logging technology that opened up timber resources in wet environments whose sogginess made them inaccessible to trains, mules, oxen, and carts. The methods that loggers had used for decades in the swamps, bogs, and bays—girdling and felling logs from a boat and poling them to a collecting pond—were not sufficient to meet the growing demand for timber. Horace Butters, a lumberman from Michigan, thereupon decided to experiment with his overhead logging cableway steam skidder, for which the U.S. government had granted him a patent in 1883. He mounted it on a scow and floated it into a Columbus County swamp, where timber workers used its steel cables to pull cypress trees from the swamp. Because of the cable's initial length of 700 to 800 feet, this new method opened up only a limited range of merchantable timber. It did, however, accelerate the logging of previously untapped timber resources. Once the technology had been improved, pull-boats could skid logs for up to 3,000 feet.[42] Further technological strides only heightened the industry's impact on the forest, increasing the number of tree species it could profitably process, as well as the speed with which it could do so. The development of the dry kiln turned previously shunned smaller loblolly pines into desirable trees for logging and "stocks of all sizes of this pine came into general demand for lumber purposes";[43] steam-powered circular saws and band saws provided a continuous, double sawing motion that helped sawmill operators crank out lumber and reduce the waste inherent in pit-sawing;[44] and the *Lumberton Argus* marveled at the Butters Lumber Company's "powerful dredge, with its great steel dipper, eating its way through the swamp, uprooting huge stumps as easily as a dentist pulls a tooth, . . . and drawing behind it a house boat, in which the crew cook, eat, and sleep."[45]

These technological advances, combined with such mobility of workers and machinery, facilitated the construction of moveable camps and mills that enabled companies to cut cypress, cedar, gum, and other swamp species deeper into swamps than previously possible. The Butters Lumber Company, for instance, operating from Boardman, Columbus County, on the eastern bank of the Lumber River and on the edge of the Big Swamp, could produce daily 100,000 board feet of North Carolina pine, cypress, and black gum.[46] Now able to cut timber deeper into the forest and move processing machinery closer to fresh stands, lumbermen saw fit to house their workers closer to logging sites. The need to operate all this new technology on a continuous basis in order to make larger pools of fixed capital worth the investment spurred lumber companies to abandon the traditional practice of hiring seasonal labor from farming communities;instead they housed a permanent labor force close to their mills and logging sites. Building barracks, providing food, and offering limited services, companies erected camps and mill towns in an effort to maintain a steady work force in an industry growing notorious for low wages, hazardous working conditions, and poor treatment of workers.

While logging camps were legendary for their frontier atmosphere of violence and vice (one observer called them "on the whole demoralizing, for liquor there flows freely"[47]), the camps were really outgrowths of nineteenth-century industrialization: concentrations of labor power in the vicinity of fixed capital where workers put their lives and limbs on the line for meager wages to the much greater benefit of capitalist enterprises. The mechanization and capitalization of the lumber industry greatly limited the kind of freedom John Baldwin enjoyed.[48] Although workers were relatively free to roam from one logging camp to the next, once in the employ of a lumber company they were subject to the kind of working hazards and stringent supervision that Baldwin had not experienced.[49] The North Carolina folktale of a saw blade so sharp it kept going for miles after it came off a running mill most likely originated among men who lost their limbs, lives, and colleagues to the pine forest's moveable, always buzzing saws.[50] In addition to the traditional hazards of falling trees and working with stubborn mules, lumber workers faced flying timber, ravaging fires, exploding boilers, and relentless machines. In one instance, a young man was "caught by the log carriage and drawn into the swiftly revolving saw, his head . . . split in half and his body . . . mangled"; in another, a fireman was "blown to bits" by a "bursting boiler."[51]

Without this level of command over labor, capital, and technology, smaller local companies like the Southport Lumber Company, incorporated in 1890 with a capital stock of $10,000 and commended for its contribution to local economic development and social life, were unable to take advantage of the area's natural resources. The wood the Southport Lumber Company handled best was that of the bats its team of employees swung in local baseball games. For the timber it needed to feed its sawmill the company depended on local timber owners and on transports from Wilmington, which did little to help the company supply the local market with affordable sawed lumber, and it petered out quickly.[52] For a while after the company's collapse, the only wood available locally was that produced domestically in small quantities by local citizens.[53] When Southport's lumber industry reawakened it emerged in a form structurally different from that of the older, local company, a development with far-reaching consequences for the southeastern corner's economic and environmental development. In 1915 the town welcomed a new incarnation of the Southport Lumber Company, but this time the company's proprietors hailed from outside the region, owned the timberlands with which they supplied their sawmills, and exported the lumber they produced. The Benjamin-Graham Company of Salisbury, Maryland, bought a large tract of timberland, erected two sawmills, and employed about thirty men to cut timber for shipment to New York and Philadelphia. A 1917 fire "totally destroyed" the company's mill plant and ended its operations after less than two years, but the fact that $10,000 and 500,000 feet of lumber went up in flames indicates that the second coming of the Southport Lumber Company had a far greater economic and environmental impact than the first.[54]

The growth of the lumber industry in Columbus County presented an even greater intrusion in the rural economy and the environment around the Green Swamp. Outside companies such as the Cape Fear Building Company, Waccamaw Lumber Company, and the North Carolina Lumber Company (founded by New Yorker Charles Beers) tended to carve out large spaces for themselves on the margins of local economies. They drew local and foreign labor into the heavily mechanized workforce of sawmill towns and logging camps newly erected on the edge of company-owned forests. The timber they harvested usually served distant markets, and the companies were only marginally integrated into the local economy. The Cape Fear Building Company, for example, was founded by "Northern men, carpetbaggers," whose production of sash blinds and frame houses for the national market generated annual net profits of $20,000. By 1875 the company's largely self-contained plant boasted sawmill machinery and tools, factory buildings on 500 acres of land, factory machinery and tools, 4,000 acres of timberland, and mules and wagons, altogether worth over $100,000. The company drew labor from the region, but did not depend on local manufacturers and business for any auxiliary services.[55]

Operating from Bolton but incorporated in Michigan with $1 million in capital stock, the Waccamaw Lumber Company was a full-fledged capi-

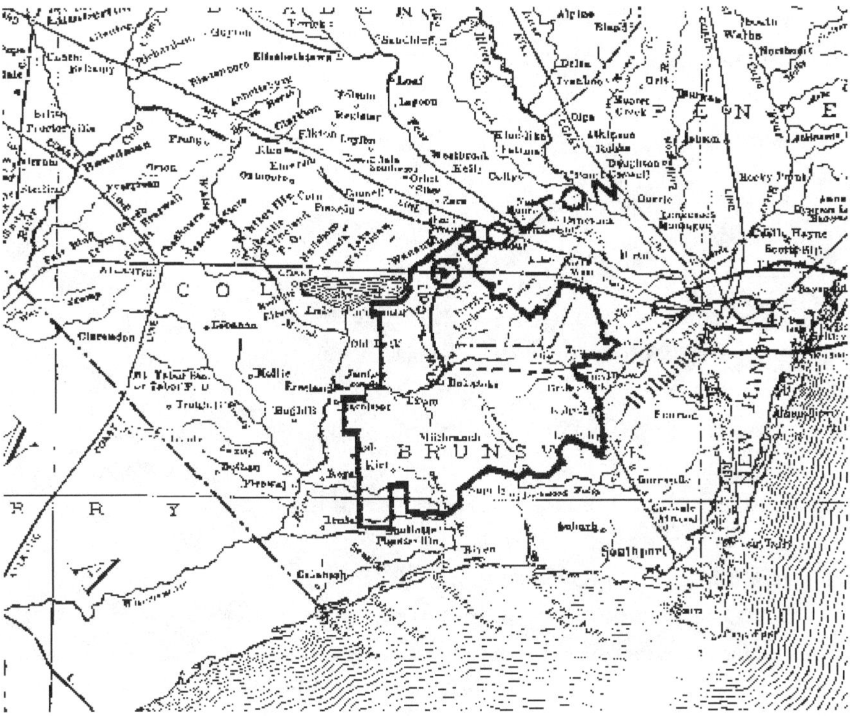

Figure 5. Holdings of the Waccamaw Lumber Company outlined on a map of southeastern North Carolina. (Courtesy of Hugh MacRae Papers, Duke University.)

talist enterprise that brought to bear on the forest environment the labor organization and steam-powered tools of modern industry. In 1906 it bought for one dollar the former property of the Green Swamp Company, the immense tract of rugged land that had since passed into the hands of the Waccamaw Land and Lumber Company.[56] Over the next six years it added to its holdings through land purchases from individuals in Columbus, Brunswick, and Bladen until it owned over 200,000 acres of heavily timbered forest- and swampland worth $750,000.[57] The company, which also operated a plant in Wilmington, built its Bolton mill close to the Atlantic Coast Line and constructed two branches that converged in the heart of the plant; it owned twenty-five miles of railroad, three locomotives, and fifty log cars, together worth $37,289.[58] At the plant, a large boiler provided steam power for the sawmill, the shingle mill, the planing mill, and the box mill, the heavy machinery for which came from the Prescott Company of Menominee, Michigan. Besides various mills and lumber yards, the plant also housed a log pond, a water reservoir, several

Figure 6. Railroad building in the Green Swamp, early 1920s. (Courtesy of Waccamaw Lumber Company Collection, Duke University.)

water tanks, a trash burner, a large storage shed for shingles, a filing room, a dust house, and a repair shop. The company owned $107,980 worth of mill and kiln buildings and machinery, which by 1915 enabled the company to produce North Carolina pine, cypress, cedar, poplar and tupelo boards at a daily capacity of 100,000 board feet.[59] When fire destroyed the company's mill, boiler room and electric plant in 1916, the financial loss amounted to $190,000.[60]

The diversification of the company's plant made it a largely self-contained operation that facilitated the stripping of unprecedented tracts of surrounding forest. Command of machinery and labor also enabled the company to make deep inroads into the largely untouched forestlands they owned, in effect moving their logging operations away from town centers and isolating their activities from regional markets. The Waccamaw Lumber Company owned five skidders and two loaders worth $16,492 and used them to penetrate the dense forests and swamps with the hard-fought construction of logging railroads—"spurs"—on the soggy soil. Once the railroad was in place, logging crews moved in with their mule-drawn carts and steam-powered skidders and harvested the timber within an area about 500 yards around the spur track. They felled trees and "snaked" them out of the woods and toward the railroad track, where a steam crane loaded them onto trains. The trains would then take the logs to the sawmill or the log pond. Once a certain area no longer yielded merchantable timber, the workers tore up the track and moved it to fresh timber stands, which ultimately resulted in a stretch of barren land scarred with cribs, snaking tracks, stumps, and the refuse of a temporary human presence. When the company put up for sale part of the land it had logged, even the West Virginia Pulp & Paper Company, whose standards for harvestable timber were far less stringent than those of lumber producers, found the land too depleted for a pulpwood harvest.[61]

Equally disconcerting, the bulk of the lumber the company produced in the southeastern corner went to markets outside the region, a reallocation of goods and resources that amplified the company's transient character. The exact destination of the Waccamaw Lumber Company's products is unknown, but northern lumber companies of similar size operating in the South tended to ship their lumber either to urban yards in the North or to markets in South and Central America, the Caribbean, and Europe. While

Figure 7. Mechanized loading in the Green Swamp, early 1920s. (Courtesy of Waccamaw Lumber Company Collection, Duke University.)

ports along the Gulf Coast had supplanted Wilmington and Southport as the main shipping points for lumber to the West Indies, New York City retained its importance as a destination for the lumber that larger companies—such as the Angola Lumber Company and the Kidder Lumber Company—produced in southeastern North Carolina. Other large producers sent their loads to Hamburg, Germany, where walnut, poplar, cedar, gum, and pine from the southeastern corner were used for the manufacture of pianos, automobiles, wagons, and army equipment.[62] Although such Atlantic trade connections constituted temporary economic boons, they mostly sustained an exodus of natural resources that in the long run altered for the worse the region's natural resource base without producing sustainable long-run benefits in return.

Perhaps the company's activities would have been more fully integrated into the local economy had its workers dispersed to their homes and farms after a few days' work as part of the exchange between farm and forest that characterized life on the Lower Cape Fear. As it was, the labor force was concentrated in Makatoka, the company's logging camp in Brunswick County, and in 125 cottages in Bolton. Like all lumber camps, remote little Makatoka "bore the reputation of being the roughest place in the South." Of greater importance to the way larger lumber companies isolated their interaction with the forest environment from local practices, however, was the fact that, as one former employee put it, "Waccamaw Lumber Company owned everything in sight." The company's assets included all the facilities of the Makatoka camp as well as practically all houses and lots in Bolton. It only increased its economic autonomy when it attempted to secure permanent occupancy of this $53,371 investment by housing native and foreign labor. While on the one hand the company followed local practice when it housed black workers in segregated areas and hired them almost exclusively for menial positions, on the other hand, it affirmed its "outsider" status by hiring foreign workers who did not grow roots in the community but stayed just long enough to make locals notice their foreignness.[63] In 1910 three Finnish households—two made up of families who had come to North Carolina by way of South Carolina and Georgia and the third made up of three single men—added a Finnish flavor to the camp. Locals also remembered Italians, Poles, Swedes, Russians, and Hungarians who did not mix with the natives: "Them foreigners camped to themselves." One worker recalled over 10 Italian families forming "an Italian City . . . at Honey Island Swamp. They was a lot of them." Only two

Figure 8. Lumber workers posing with guns and liquor at Makatoka Lumber Camp, early 1920s. (Courtesy of Waccamaw Lumber Company Collection, Duke University.)

Italian men were exceptions, marrying local women and starting families there. Having found the Waccamaw Lumber Company at best a temporary boon, the other foreigners—at least "those that didn't get killed"—proved as transient as the company and left the communities around the Green Swamp to deal with the environmental legacy of their labor.[64]

For lumbermen, after all, this newly achieved mobility of labor and capital constituted a victory over the environmental obstacles to economic growth that the longleaf pine forest had long posed. Freed from dependence on waterways, they harvested timber almost anywhere and transported it all over; with more timber stands and more tree species than before accessible for profitable exploitation, lumber companies went down to work. "Now clear cutting is the rule in lumbering," North Carolina forester W. W. Ashe warned in 1894, "though much to the detriment of the forest."[65] In 1893 alone 28 of the lumber establishments operating in the southeastern corner manufactured 20.84 million board feet of lumber and 22.6 million shingles.[66] In 1896, 20 million board feet of longleaf pine lumber left the port of Wilmington.[67] As a result, the industry wrought dramatic changes in the forest landscape across the Lower Cape Fear region.

Trees disappeared by the millions, sometimes replaced by inferior species; the extensive snaking and dragging of logs across the forest floor caused severe soil erosion; the disappearance of trees from the banks of rivers and creeks destabilized wet environments, while the continued rafting of logs caused waterways to clog and flood; and fires ignited by sparks from trains burned large tracts of forest and added to an overall decrease in timberland and wildlife habitat. Lumbermen, however, acknowledged the downside of the transformation they effected only to the extent that it affected their business, and usually responded by moving, not changing, their business.

"We have cut only a little timber"

Charges of "assault" and "plunder" are not unwarranted, then, in light of the lumber industry's cut-and-run response to the environmental havoc it wreaked; the conceptualization of the industry's activities as a "capitalist onslaught" nevertheless obscures the nature of economic and environmental transformation as it took shape in older rural communities. In and around the Green Swamp, the harbingers of modern industrial capitalism set up shop in a rural economy where the demands of farm, family, community, and living in close proximity to swamps and forests did not so much contradict as mitigate a shared desire for economic development. Unlike the lumbering giants that took advantage of the wide availability of cheap federal lands to accumulate and denude uninterrupted expanses of pine forest in Florida, Louisiana, Texas, and Arkansas, quickly closing the door of opportunity on smaller competitors, large companies in southeastern North Carolina, for the most part controlled by outside capitalists, coexisted with smaller, locally owned businesses that exercised more care to fit their activities into the existing patchwork of farms, forests, turpentine orchards, swamps, and railroads. Utilizing crude technology, employing local labor, and depending on local timber owners who had not sold out to larger corporations and with whom they often forged relationships in their capacity as general merchants, local entrepreneurs had a less destructive impact on the environment and a more constructive impact on the local economy. The existence of larger numbers of small companies reflected the prominence of widely distributed landownership and the more fragmented, less sweeping character of logging practices. The inroads these companies made into the forest environment, as well as the different relationships they forged with surrounding communities, reflected the kinds and grades of machinery they utilized, the size and makeup of the labor forces they employed, and the level of control they exerted over the timber stands they harvested. The ecological and social impact of "the lumber industry," although nowhere negligible, depended on the degree of coexistence of large outside corporations with smaller local companies, which created a patchwork of logging and sawing sites where lumbermen transformed both environment and community at different levels of intensity.

In Hallsboro, a few miles west of Bolton, Pierce & Company decided in 1913 to add to its mercantile business "a small saw-mill, and possibly a little planing machine . . . to be run by the [cotton] gin engine."[68] Over the next seven years the company, whose diversified business was typical of successful local entrepreneurs in the postwar South, purchased both new and used machinery to set up its operations and to replace material losses inflicted by a fire. By 1920 Pierce & Company operated two sawmill plants, one of which included a tram road valued at $580, as well as a moveable sawmill. In terms of capital, the company certainly was not a lumbering powerhouse—the 1917 fire, which shut down the sawmill and the cotton gin, did only $2,211 worth of damage. But with two boilers, a sawmill, a planing mill and building, two engines, a dry kiln, and a lumber shed, the company's Mill Plant No. 1 was a complete lumbering outfit: in 1918, it sawed 918,035 board feet. With one boiler, one engine, a sawmill, a trimmer, an edger, a cut-off saw, and a tram road, Mill Plant No. 2 was a less intricate affair, yet it still managed to saw 579,493 feet in 1918.[69]

Because of Pierce & Company's limited timberland ownership, the operation of its sawmills directly benefited the local economy, as the company contracted with absentee and local landowners for a steady timber supply while harvested lands remained in the hands of owners who tended to see them not as a burden but as an opportunity. Many of the company's dealings were with former Hallsboro resident D. M. Flynn, vice president of Flynn-Harris-Bullard, a naval stores and wholesale groceries business in Florida and Georgia. Flynn still owned several tracts of land around Hallsboro and agreed to sell Pierce & Company the timber. In April of 1916, for example, Flynn sold to the company, at a rate of $2.50 per thousand feet, "all the pine timber, hickory, gum, oak, ash or any other standing timber" on a stretch of land south of the railroad. He planned to divide this particular tract into ten-acre farms that he could sell to prospective settlers and depended on Pierce & Company "to cut over this timber closely" before 1 December of that same year.[70] Resident landowners, too, supplied Pierce & Company with wood through timber deeds, often selling the pine stand on a few of the acres they owned. In 1915, for example, Flora George sold Pierce & Company all timber ten or more inches in diameter on three tracts that had become hers upon her husband's death. The company promised to cut the timber within seven years, paying George $10, $5, and $1 per thousand feet, depending on the logs' diameter. In 1919 Ben Mitchell and his wife signed a five-year deed that gave them $62 for all the pine timber at least ten inches in diameter on their thirty-five acre home tract.[71]

The logging and sawing of timber also facilitated a range of different arrangements between the company on the one hand and landowners, timber owners, farmers, and workers on the other. Because of the relatively small scale of its operations, the company could not afford to hire a pool of steady loggers and house them in lumber camps. Instead, they hired local labor for logging jobs on an individual basis. In one instance, they paid a black man named Frank Williams for "getting out shingles" on lands near Wananish, the owner of which had deeded the timber to Pierce & Company.[72] In another instance, George Frink agreed both to sell to Pierce & Company "all the pine timber twelve inches or larger at the stump" on two tracts of land and "to cut and haul the pine timber from [these] lands . . . to Pierce & Company's mills." The company furnished a sawmill and sawyers for each tract, but Frink did the logging and hauling.[73] In a third deal, Pierce & Company hired W. L. Burchett "to cut and saw into lumber" the timber on a tract of land for which the company held the timber rights. In this case, Burchett furnished the sawmill, hired and paid the labor, and provided for "repairs, supplies and upkeep." Pierce & Company paid him $8.50 per thousand board feet of lumber, sawed to size and delivered to the company.[74] At other times the company bought timberland, only to sell it again while reserving the timber rights.[75]

The use of local timber owners in these various capacities helped Pierce & Company maintain a fairly steady supply of timber and lumber for longer periods of time: from 12 September 1930 to 4 September 1931, for example, the company paid $25,988.65 for fifty-two weekly lumber purchases ("total lumber bought") that amounted to a total of 1,951,850 board feet.[76] Not only were the relatively small amounts of lumber bought and money spent indicative of the scale of this local operation; while the average weekly purchase was some 37,500 board feet, the company bought only a total of 10,000 board feet during three weeks around harvest time, which suggests its dependence on local farmers for timber and labor. On nineteen dates, the company paid J. K. Baldwin for his contribution to the weekly lumber supply; the company paid him for his first ten loads on 12 September and on the following nine paydays, but his last load did not come until 30 June. His sporadic sales earned him $559.63 for a total of 43,831 board feet; and even if he also sold lumber to other companies or individuals, it is likely that he logged his land occasionally until he exhausted his timber or no longer considered harvesting trees necessary or viable. During the same year, Pierce & Company credited E. J. Clifton $1,511.18 for 123,421 board feet of lumber, which he supplied in 27 shipments. Unlike Baldwin, Clifton clearly tried to clear a piece of land as quickly as possible: the company paid him for his first twenty-six loads on continuous paydays and the last following a two-week hiatus, after which he sold no more. These different practices by a number of farmers and workers served different economic purposes that had a much more diffuse impact on the forests of Columbus County than large-scale lumbering, yet added up to a steady supply of lumber.[77]

Moreover, given the nature of these exchanges, as well as the time it often took Pierce & Company to harvest what were relatively small tracts, landowners did not face rapidly and violently denuded lands but gradually thinned out woodlands that could regenerate more easily. The logging and sawing of these fairly small timber stands often presented quite a challenge without the heavy machinery and large labor forces available to large companies, resulting in stretches of timberland left untouched for extended periods. "The weather has been so wet that we have cut only a little timber . . . for several weeks," the company told D. M. Flynn in March of 1917. In September the company could not "get the timber out of the branches and swamp except in dry weather." During the intervening six months, the company at one point sawed only 677 board feet in a week, 60,055 board feet in two weeks, then 78,379 board feet in a month.[78] In addition, of the twenty-eight timber deeds recorded in one of the company's ledgers, only one restricted cutting to trees with a diameter of eight inches or more; the other deeds that came with a diameter specification all drew the line at ten or twelve inches. This was an economic decision with both short-term and long-term dimensions: smaller trees did not produce much lumber, so the cost of logging and sawing easily exceeded the potential return; allowing smaller trees to grow guaranteed a future supply. In at least one case, moreover, Pierce & Company left standing two longleaf pines that were eligible for logging to re-seed the land.[79] As a result, Pierce & Company had a smaller physical impact on, yet deeper roots in, the environment and community in which it operated than a corporation like the Waccamaw Lumber Company, which simply sold its land when the timber gave out.[80]

Pierce & Company's employment of workers from the local labor market further impeded the kind of rapid logging that larger companies were able to sustain and enhanced the ties between the Hallsboro community and the local sawmill. The company did not own the kind of large operation that necessitated the continuous employment of a large group of workers to make it pay. Indeed, working for Pierce & Company was less strenuous, rigid, and specialized than working for the Waccamaw Lumber Company. Many workers worked only a few days per month, some as little as half a day in the course of two weeks. Some worked as much as twelve days in two weeks, but most employment fell somewhere in between. In addition, workers were not limited to one task: workers drove the mules one day, a truck the next, and worked in the planing mill the following week. They also were paid at different rates; not only did men contract for different hourly wages for the same work, some of the loggers were paid by the 1,000 board feet they produced while the mill crew were paid by the hour. Workers sometimes could earn some extra money by stepping in when needed unexpectedly, as when they helped put out a fire at the planer or spent six days "cleaning up." Finally, while many workers were typically paid in store credit, some received cash for their services.[81] The conditions under which Pierce & Company and its employees labored, therefore, are much more consistent with those of the lumber industry as it operated prior to 1880.

The more fluid nature of Pierce & Company's operations is indicative of its being rooted in the social and eco-cultural life of the Hallsboro community, and reflected the variegated economic activities of areas that did not have a large and concentrated labor force working in one industry. Whereas towns such as Bolton and Boardman were home to large numbers of lumber workers and households that shaped their economic life to accommodate and profit from their presence, households in and around towns like Hallsboro tended to be of a mixed nature. Barter trade was still common before 1940, peddlers took their wares from town to town, and boarding houses often bought produce directly from farmers.[82] Heads of households continued to farm while their older sons or live-in relatives took jobs at the sawmill; in other cases, wives tended to the farm, sometimes on their own account, while their husbands joined their offspring in the mills or on the railroad.[83] Pierce & Company often bought timber from local farmers and then employed those farmers as the loggers and sometimes as the sawyers of their own timber stands.

These were people who, unlike the workers corralled into company settlements and dependent on company credit, went home after a day's work, and Pierce & Company saw itself as part of that community more than as part of the national lumber market and business world. J. B. Wyche expressed to D. M. Flynn his concern that "the larger part of the pine timber in this section has been bought up by outside parties, and much of it is being cut and removed to be manufactured elsewhere, giving comparatively little work to the people of this community."[84] Pierce & Company took pride in the fact that local farmers asked it to "saw logs . . . for their own use," that people purchased plane mill ends (knots, splits, and defects that were cut out of the timber) for firewood and cooking stoves, and that customers paid accounts with timber, lumber, or logging work.[85] The muscle developed in one of the company's sawmills by "Big Jim" Thompson, captain of the Duke University football team and son of one of the company's partners, was a source of community pride.[86] And even though a large proportion of Hallsboro's farming community was indebted to the company through crop liens and mortgages, the company considered itself to be "not very pushy about collecting as much as the rest of them."[87]

"For the good of all concerned"

If these actions and attitudes seem paternalistic, it is because they are: self-professed corporate benevolence is rarely intended as an expression of "solidarity with peasants," and we should not treat it as if it were; neither is such discourse of "community" simply so much corporate hubris masking cold economic calculations. Pierce & Company was part of a larger network of businessmen who, like its predecessors during North Carolina's naval stores boom, saw its efforts as proper and essential to the development of its region and sought to impose its vision of improvement and legibility on landscape and community alike.[88] Moreover, these values were not by definition anathema to those of "the community" or "the people"—the latter were not the guardians of nature against the ravages of business. Although many towns that sprang up during the lumber boom owed their existence and their names to entrepreneurs, only R. R. Barnes of Barnesville seems to have earned a reputation as his town's "dictator"—and even that moniker reflected admiration as much as disdain.[89] The ventures of James H. Chadbourn & Company, former shipbuilders and successful lumber traders in Wilmington since before the Civil War, suggest the power entrepreneurs wielded around the Green Swamp.[90] The Chadbourn Mill and Railroad Company, incorporated in 1887 with several Chadbourns as primary stockholders, connected Chadbourn (formerly Timberville) on the Wilmington, Columbia & Augusta with the timber stands along the Waccamaw River in Horry County, South Carolina, to haul timber north to the W. C. & A. rather than raft it downstream. The Wilmington, Chadbourn & Conway Railroad turned the little hamlet into a bustling lumber hub.[91] James Chadbourn Jr., was president of the Wilmington Chamber of Commerce in 1899, an influential figure to whom timber inspectors turned for suggestions when they needed to hire a replacement.[92]

The Chadbourns shared their vision of progress with the likes of Charles Beers Gault, who urged a congressman to pursue the building of a road between Whiteville and the Green Swamp. It would be "for the good of all concerned and especially for the development of our county," he wrote, while "a couple of years of delay in the development of Columbus County . . . would be anything but progressive."[93] When Gault and Charles Oscar Beers wished to sell some land, they praised not only its proximity to the railroad and to Lake Waccamaw, but also its suitability for growing crops *and* the great opportunities the property afforded for hunting, fishing, and boating.[94] Like so many of their peers, the Gaults and Beers, while engaged in forest devastation, could still wax poetic about the region's nature: "There are so many pleasant walks about here," wrote one of the clan. "The lake is always beautiful and the walk up or down the shore delightful. Then up the road in one direction is a lovely piece of woods . . . and down the road is a large pine grove, where in places the myrtle grows very thick." These were not contradictory ideas: pastoral nature was to be enjoyed, while places where "the gray moss is so abundant with its *ghostly* festoons as to give you *uneasy sensations*" demanded man's intervention.[95]

Another business well-connected to community and land was the Council Tool Company of Wananish, a manufacturer of turpentine tools. Its founder, John Pickett Council, nurtured attitudes toward nature and commerce that reflected those of the communities around him. Council was a farmer, woodsman, and turpentine producer who understood the importance to the rural economy of multiple, coexisting forest uses, and whose interest as a manufacturer of forest exploitation tools was in a sustained supply of pine stands. His love of fishing was so well known that an 1899 report of the company's move to Lake Waccamaw prompted the observation, years later, that "Council could now fish to heart's content, and be only a mile from his tool business." After his death, the town of Whiteville passed a unanimous resolution encouraging the North Carolina Shipbuilding Corporation to name a vessel for him: "The late John Pickett Council was a great and outstanding contributor to the state . . . industrially and pioneered in the movement for the preservation of game, fish, and wild life . . . He . . . contributed no little to the welfare of his state and community."[96] In a region where farmers, laborers, and businessmen alike valued hunting, fishing, and the use and enjoyment of land, businessmen such as Council were far more likely to be seen as role models than as dictators.

Perhaps the best indication of the intertwined nature of business and community values is the modest correspondence and diary of Samuel James Rowell, a farmer and small-scale lumberman in Brunswick County. Rowell and his family grew a variety of crops and kept several farm animals, mostly for domestic consumption. Like so many farmers, they recorded such natural occurrences and environmental interactions as the success of crops; their work at hoeing, plowing, and planting swampland; the rising water in the creeks; their dread of hog-killing time; and the loss of chickens to minks. But Samuel Rowell also harvested trees and rafted them downstream to Wilmington; and during one such logging and rafting trip in 1907, he kept a diary for his family in which he revealed the worldview of many enterprising men around the Green Swamp. On one day Rowell noted the beauty and stillness of nature, on another the disruptive force of wind and rain; one night he wished his wife were with him when he killed a squirrel, "butchered it," then boiled it and "fried it Brown and ate it with potatoes"; he noted the smoke of a fire near the Green Swamp; he recorded building a horse shed and a chicken coop in his logging camp, as well as the day the hens commenced laying; after a good day's work, he felt "thankful for the Blessings" he received that day; and most nights he read in his Bible, recording the passage for each night and sometimes urging his family to read them as well.

In his economic activities, his ties to the land, and his outlook on life,

Rowell was not much different from many generations of farmers before or after him. Judging from his writing, he witnessed neither a "capitalist onslaught" nor the "coming" of a New South, and he did not experience his "moral economy" as a bulwark of tradition in the face of sweeping changes. Rather, he was himself an influential and respected member in his community as well as in his church, as his practical reflections on church rituals and scripture passages suggest. The church had remained an important guardian of social order, and many of the business community filled leadership roles in congregations whose records suggest they never questioned business practices but excluded members from fellowship for "*disorderly* walk"—fornication, drinking—in their personal behavior (italics added). Modern Christianity's dogmatic separation of individual wholesomeness from profit-driven market behavior only strengthened the position of sober, economically successful men in their communities, for whom they were models of holiness and industry alike. "I am 54 years old today," Rowell wrote on 11 December 1911, "and I thank the good Lord that it is as well with me as it is. I am lonely tonight." Life could be hard and demanding, but it was good and prosperous all the same, and affirmed the values of men of his class.[97]

"Wanton Destruction, Indiscriminate Wielding"

Much of the rural population subscribed in whole or in part to these elite visions of progress, land use, and rural life; and because the advanced lumber industry and its auxiliaries built on traditional ways and means of resource extraction, the people around the Green Swamp observed them with far less hostility than our language of violent invasion and abuse would suggest. With the lumber boom came employment opportunities, more and improved transportation facilities, more and better lumber at lower prices, and an infusion of capital into the rural economy, all of which could enrich communities without necessarily pulling up their roots in the forest environment. The *Southport Standard* reported that as a result of people finding employment with the Gardner & Lacey Lumber Company in Excelsior, "There is much more money in circulation . . . and our merchants are selling more goods . . . than for some time past."[98] The appearance of local and outside companies forced locals to weigh the benefits of industrial employment, additional railroad mileage, and abundant lumber against diminished woodland access and large tracts of denuded forest.

Most people in rural communities were certainly not averse to accumulation and believed that farming naturally followed logging; but while they usually welcomed, if reservedly, the economic impetus of lumbering, they still experienced unease about outside monopolies, forest decline, and foreign labor.

Reservations toward the lumber industry arose when the presence of lumber companies, especially "outsiders," wore away the fabric of rural life, which combined the pursuit of economic development and monetary gain with a more reverential attitude toward soil and elements, as well as with access to woodlands for firewood, hunting, and other recreational and sustenance needs. The voices of protest against lumbermen around the Green Swamp were not those of powerless wage workers or mad preservationists, but those of farmers and town boosters who worried that outsiders came to play the "grab game": snatch a town's natural resources and turn them into a quick profit without investing in long-term enterprise and local development. Farmers who propagated caution and boosters who championed commerce both stressed the effects any lumbering enterprise might have on a forest that was part and parcel of rural life.

Farmers objected not to the cutting of trees for domestic or commercial purposes but to the ways in which the activities of outside lumbermen undermined local autonomy and relationships with the environment. During the Populist era, with farmers especially wary of anything that resembled a lumber trust or any other potential threat to rural life, Raleigh's *Progressive Farmer* held that "while the woodman's axe has done much good in carrying civilization to the borders, in its *thoughtless, indiscriminate wielding* it has also done much harm" (italics added). Its greatest objection was to Northerners who took advantage of local landowners by "buying immense tracts at very small prices, the object being to cut down, saw up and ship the lumber North." Yet the journal advised readers on how to "destroy" swamp willows that "infested" certain valuable lands, then urged farmers to plant trees to replace those they sold to lumber companies, and counseled farmers to set aside land for planting walnut trees to ensure their sons' economic independence. It ran an advertisement for Salem Iron Works, a regional manufacturer whose machinery affirmed the interconnectedness of different forest enterprises: it produced $150 sawmills for farmers and larger sawmills for lumbermen. The journal also included advertisements for hunting rifles, but urged an end to "the wanton destruction" of the black snake. Depicting the much-loathed serpent as "a friend to the farmer," the

Progressive Farmer affirmed the forest's importance as an indispensable part of rural life, "as a protection against the sweeping North winds, as distributing agencies of rain." Harvesting timber for money was good and desirable, farmers argued explicitly and implicitly, but that aspect of forest use should neither become the exclusive domain of outside interests nor obliterate the forest's functions as habitat, hunting ground, and environmental regulator.[99]

Champions of economic growth tended to see the enterprise of outside interests as primarily beneficial to local economies, yet they, too, saw the forest environment as a whole play a crucial role in local socioeconomic relationships. While they stressed the value of local timber stands to potential investors, they also tried to convey the natural beauty and healthfulness of the place to which they wanted to attract capital, as well as the abundance of game for recreation and the thriving farming community nearby. The *Southport Herald* lifted up the area's faunal wealth as an attraction for sportsmen, depicting Southport as an ideal home base where one could "always find a day's sport within a half mile of the town."[100] The town was one of several whose businessmen sought to attract capitalists from outside the community to develop—without compromising the blessings of life in town, farm, and forest—the potential for economic prosperity that nature had bestowed on the little port. Its surroundings abounded with woods and water, and entrepreneurs kept up a chorus of promise and apprehension that Southport would prosper as a town where the blessings of farm, forest, and ocean converged once capitalists committed to the construction of a railroad. Connecting the town to forests and to larger markets near and far, the railroad would stimulate commercial agriculture and, with coal coming in from to the Appalachian coalfields, establish Southport as a major coaling station for large steamers traveling along the Atlantic coast.[101] William Curtis asserted confidently that "the great tracts of valuable timber in the Green Swamp would . . . seek an outlet in Southport"; qualifying his blanket condemnation of the forest industry's destructiveness with a sense of local urgency, he suggested that Wilmington take part in the development of its neighbor rather than leave it to outside capitalists who cared "only for their own interests."

Implicit in this public criticism of outside interests was a condemnation of individuals who did not subscribe to the larger vision of rural development and proper resource use. Wantonly destroying black snakes, leaving willows to infest the land and the cutover barren, obstructing railroads and other enterprise, disregarding private property—these were seen as obstacles to progress, and, more often than were the exploits of capitalists, subject to local policing and prosecution. Emblematic of this worldview are the different communal responses to mutilations and deaths caused by wild animals and those caused by industrial machinery. Bears, snakes, and alligators that attacked and killed humans—or came close enough to appear threatening—were hunted down and destroyed; machines that ate people were never the target of extinction campaigns. Industrial threats to life and limb were the acceptable and inevitable dangers of progress: Charles Beers Gault witnessed one of his millworkers "holding up his hand with his *fingers* hanging down;" but when another asked to be given "the job of removing the dust from under the equalizing saws," Gault did "not consider it any more dangerous than any of the work in the mill around the machinery."[102]

The traditional practices of gathering firewood or hunting in easily accessible forests increasingly provoked the ire of businessmen committed to orderly exploitation and recreation. John McKay, who in 1872 set up shop along the Carolina Central in Shoe Heel, Robeson County, to farm, distill, and ship barrels of rosin and spirits of turpentine to Wilmington, tried with his fellow entrepreneurs to restrict easy access to what were often poorly marked private lands, forbidding "all persons . . . to hunt with dogs or guns or trespass in any way . . . under the strict penality of the law in every case."[103] A land surveyor ran into opposition on Crusoe Island, where settlers who had registered claims continued to cut timber. A civil engineer advised Francis Gault, who claimed the timber rights to the island, to get a court injunction and keep the inhabitants "from cutting any more timber and have a court surveyor appointed to run out the lands. If you don't do this, all your timber will very probably be stripped off the land before you can get a satisfactory settlement."[104] Anger at the surveyor's inability to do his work clearly shows the importance of timber surveys in the effort to impose a legible and profitable order on the landscape: not only did they reduce the landscape to a two-dimensional grid of calculable values, they also helped timber owners track environmental change in simple but crucial terms of "thousand board feet" and "dollars," of "rightful" and "illegal" use.

Proponents of "proper use" within the lumber industry saw themselves as guardians against deforestation; they tried to address the potentially hurtful consequences of forest product consumption for the future of timber supplies and sought to regulate the industry's own practices. The *Southern Lumberman* pointed an accusing finger at American consumers, who reportedly consumed 400 feet per capita per annum while Europeans needed a mere sixty feet. Americans, the *Lumberman* charged, tended to use wooden packaging materials such as boxes and crates only once and then discard or burn them. The "sixteen million feet of lumber lost in this sort of waste is a great drain on our forest resources, now none too plentiful."[105] The creation during the early twentieth century of the Southern Pine Association politically cemented the superior position the "proper" users of the business community had long claimed over the rogue exploiters of the forest, in this case the wasteful "peckerwood" mills.[106] The rapid turnover of mills in North Carolina, to be sure, was not so much a function of timber depletion as it was of small mills harvesting relatively small holdings and new mills emerging to cut the relatively quick second growth; yet standardization of lumber grades, adoption of fire prevention measures, and fledgling implementation of state-endorsed and expert-controlled scientific forestry across the South played into the hands of large, well-informed, and well-connected lumber companies, at the expense of small, local operators of no or ill-repute. Even in a state like North Carolina, especially on the eastern coast where the absence of public forestlands and the existence of more developed regional economies forced lumber companies to adjust to local patterns of landownership and environmental use, progressive associationalism among business elite strengthened the position of proponents of capitalist progress.

* * *

An 1897 survey of merchantable timber affirms that the capitalist desire to manage properly timber stands around the Green Swamp stemmed not so much from the likely disappearance of natural resources as from the many opportunities for deviant behavior nature still made possible. The existence of large stretches of merchantable timber facilitated not only legitimate resource extraction but also the kind of unlawful and roguish behavior that showed that forest and swamp were still "wild." There were still sizeable bodies of longleaf and loblolly pine in Bladen, Robeson, Brunswick, and Columbus counties, especially "along Brown and Green marshes and the flats of Lumber river and its tributaries." Bladen still had plenty of white oak, while large stands of cypress graced the swamps surrounding Lake Waccamaw and Big Swamp in Bladen. The flats around Lake Waccamaw abounded with juniper, and an extensive area of "unlumbered" white cedar forest existed in Brunswick.[107] The same survey also showed that sizeable areas of economically less desirable but aesthetically enjoyable woodland, such as the pocosins of Bladen, were pristine enough to please those who treasured nature for its recreational value. Sportsmen continued to harvest plenty of game from these lands, while others found pleasure in watching. "The closer we got to Wilmington," observed Dutch botanist Hugo de Vries in 1906, "the greener the forests were, and the richer the flora of blooming shrubs and plants." He marveled at the numerous plant and tree species in the area's savannas and swamps, including some near Lake Waccamaw that he had only seen in Dutch greenhouses: almost 200 subspecies of hawthorn, and a remarkable "concentration of insect-eating plants." After a busy week in Philadelphia, "the subtropical heat, the wonderful pine scent, and the healthy exercise" he experienced while hiking made for a delightful change of pace.

Humans had certainly altered the environment in and around the Green Swamp, much to its detriment: De Vries noted the billboards that everywhere despoiled the landscape even in "the heart of the turpentine forest's lonely wilderness." But the scars the lumber industry left on the land appear not to have dominated the landscape; in fact, the only industry De Vries mentioned was the turpentine industry.[108] In the case of cutover lands that did exist, moreover, new-growth forest on the relatively small patches lumber companies left in their wake offset to a degree the large-scale disturbance of longleaf pine ecosystems as new pines quickly replaced the old longleaf stand.[109] As late as 1925, five years after the end of the great boom, there was enough merchantable timber in Brunswick and Columbus counties to entice the Jackson Brothers Lumber Company, one of North Carolina's largest lumber companies, to set up shop in the town of Brunswick—much to the excitement of the local population.[110] When "the boys" working at the *Lumberton Argus* affectionately referred to their printing press' first gasoline engine as the "saw-mill," the purchase of which forced the retirement of the hand-press' operator, they gave voice to the industry's central place in their rural community as well as to a level of comfort with the way its capital and technological strength were imposing its leadership's capitalist vision of legibility on its still rugged landscape.[111]

4

"After the Lumbermen, What?"

Whenever a lumber company ended its operations at a certain location, it either dissolved or moved on to fresh timber stands. Either way, once the sawdust settled, a tract of cutover land came into view, an open space littered with small trees, stumps, tree limbs, sprouting brush, and surface scars from railroad tracks and snaking routes. "The timber owners of America have wasted . . . one of the most splendid gifts of nature," wrote one critical observer who accused lumbermen of "gradually making deserts."[1] Empty though it seemed, the cutover still fueled fires, sparked by lightning and fanned by unrestrained winds, that posed severe dangers to adjacent timber stands and buildings while diminishing the fertility of the soil. If left alone, it might support a second-growth forest, likely to be different from the original stand, but providing timber nonetheless. But until such second-growth trees had matured into harvestable stands, the cutover remained a scar on the landscape, a potential hazard to standing timber, and a burden on landowners who had to pay property taxes on unproductive land. The problem was such, said O. H. L. Wernicke, vice president of the Pensacola Tar and Turpentine Company and a private timber owner, that development of cutover land, the industry's "one great by-product permanent in character and indestructible in its nature," had become a dire necessity. "After the lumbermen, what?" was how he put it in a 1918 industry pamphlet. It was a query heard all too often.[2]

From the vantage point of tradition, the answer was easy. According to the land use ethic that informed the exploitation of natural resources around the Green Swamp as it did in much of the world, agriculture in the shape of livestock grazing and crop farming logically and necessarily followed the clearing of forestland. Whether it was the first batch of cotton from Columbus County to appear on the Fayetteville market in 1828, or a farmer's decision in 1894 to clear twenty-three acres of bear habitat on the edge of the Green Swamp, observers had always nodded their approval at efforts to wrest farms from the woods.[3] The patchwork of plantations and farms that had developed along the region's main waterways, pushing inland as its population grew, would have struck most Americans as the proper result of humans carrying out their duty to take full advantage of nature's bounty. "The farmer is the back bone of every community," the *Southport Standard* wrote in 1897: "without him none of us could live." The paper's evocation of God as the farmer's "partner to success," combined with its admonition to "Honor the Lord" so "thy barns be filled with plenty," reflected the belief that farming was a near-sacred vocation. Subduing the earth and turning woods into fertile gardens of seed-bearing plants and fruits, the farmers around the Green Swamp carried out the Lord's work; so the *Standard* advised them to plant various fruits and "see what God will do for you."[4] The expectation was widespread that farmers could turn the Green Swamp region's cutover into a "garden spot."[5] The ongoing "transition from primitive conditions in a great forest and sparse rural population makes elbow room . . . and means opportunity," broadcast Allen Maull of the Atlantic Coast Line.[6]

When Maull wrote of elbow room and opportunity, however, he did not so much describe the general experience of native farmers on the southern cutover as try to convince farmers from outside the region to "COME AND LOOK OVER THIS GLORIOUS LAND" and seek their fortune on eastern North Carolina's denuded forestlands. The southern cutover was so immense that the region's sparse rural population would never be able to settle and cultivate it all. On foot, O. H. L. Wernicke evoked the vastness, it would take forty-three years to travel across the cutover; by car, it would take nine; even by air it would still take ten months to fly over every one of the estimated 200 million acres of former forestland that lumbermen had cleared of trees. And this was only the land that lay idle at the time of his writing: continued lumbering activity added 15 million acres of cutover each year.[7] In addition, southern agriculture was thought to be in such a deplorable state that moving a sparse population of poor, ignorant, sickly, and wasteful cash-crop farmers onto the cutover would do as little good as leaving the land fallow. Consequently, the bulk of the cutover stood to remain devoid of further exploitation, a massive resource whose chronic under-utilization clashed with dominant views of nature as a gift for humans properly to exploit, unless a substantial number of immigrant farmers could be lured south.

From the 1890s through the 1920s a number of entrepreneurs mobilized their financial and social capital in an attempt to bring about such a demographic shift and correct the waste of resources. They sought to spark a triple revolution in southern agriculture and rural life when they developed plans to encourage white immigrants to come to the South and settle in organized farming communities on cutover land to engage in intensive, diversified, and scientific farming. First, a shortage of "dependable" white labor and the widespread dependency on "unreliable" black labor constituted a "race problem" that a large influx of white workers could help solve, especially in conjunction with the then accelerating political disfranchisement and social subordination of blacks. Second, the even more ubiquitous dependency on increasingly unprofitable staple crops on the part of a technologically backward agricultural sector would end as farmers used scientific methods to produce a wide range of previously marginal crops without wearing out their land. Third, farm tenancy would no longer impoverish farmers and erode rural life as small farm ownership in the new communities would restore the proper relationship between land and labor and the small farm as the bulwark of republican independence.[8] Like their southern, national, and international counterparts in what unfolded as a global movement, entrepreneurs around the Green Swamp saw in the establishment of such communities the solution to the closely intertwined problems of labor, race, and resource utilization—problems they considered prohibitive of proper socioeconomic development.

The ties between real farmers and their land thus obscured by grand visions of "racial improvement," "modernization," and "independence," proponents of agricultural reform failed to realize that in heavily wooded areas like the Green Swamp region their plans clashed with the relationship that had traditionally existed between farms and forests. Ties between farming and lumbering continued to extend beyond the cultivation of crops on cleared land: farmers did not share what Richard White has called "the old belief that the forest only represented an obstacle to farming," nor did they view the development of the cutover in Vernon Carstensen's juxtapositional terms of farms *or* forests.[9] Forestland surrounded farms, farms encompassed forestland, and farmers had always relied on access to woodland resources, putting the most stock in farms that included at least some timberland. In 1880 their woodland holdings covered more than half the land area of Bladen, Brunswick, Columbus, and Robeson counties. When in 1930 their remaining timber stands covered only one-fifth of the same land area, the rest likely having been reduced to more cutover land than they could cultivate, farmers were also most keenly aware of the opportunities lost in the process of environmental change. Given the continued diversity of their practices as forest users and woodland owners, they remained a check on lumbermen's single-minded pursuit of throughput maximization.

Advocates of reform, however, considered this widespread woodland ownership among farmers to be part of the problem of resource "waste" and saw in the cutover and the industry that created it the solution to, rather than a cause of, agricultural decline. Many among them lumbermen or businessmen with interests in lumbering, they wanted to reshape farming in the image of a modern business like the lumber industry. Even as their rhetoric harked back to Jeffersonian ideals of independent landowners as the republic's sturdy backbone, they championed the technology, science, management methods, and marketing strategies that had been central to the rise of industrial capitalism. Their goal was to make land and labor legible, productive, and profitable according to the capitalist definitions of those terms. In their view of proper development, the disjoining of crop cultivation on the one hand and forest ownership and use on the other, rather than being detrimental to farming, served that larger purpose. They sought to model the new southern farmer after modern capitalist entrepreneurs like themselves: a specialized small business owner, rooted far less in the environment of forest and swamp than in family and community life—according to *their* ethnic, social, and economic definitions of community.

"Deluded suckers"

Widespread disdain for "carpetbaggers" in the years following the Civil War notwithstanding, efforts to lure people from outside the South to the region's natural riches were as old as the so-called New South Creed, the belief that a South without slavery could and should follow in the footsteps of the industrial North.[10] The abolition of slavery had obliterated the region's single largest pool of working capital and transformed its labor market from one centered on high-priced slave labor into one geared toward keeping wage labor cheap. Entrepreneurs willing to operate in a biracial labor market stood to benefit from the consequences of decades of heavy investment in human capital rather than in natural resource exploitation and industrial technology.[11] Boosters were quick to present the southern economy's combination of entrepreneurial and capital dearth on the one hand and its wealth in untapped but exploitable resources on the other as a golden opportunity for enterprising adventurers. A publication that served to boost northern interest in southern economic development, *Where to Emigrate and Why* (1869) listed land, produce, minerals, game and fish, manufactures, timber, and the good people of North Carolina among the many resources that guaranteed the hardworking immigrant economic success.[12] That same year, the North Carolina Land Company published *A Guide to Capitalists and Emigrants* to persuade northern capitalists to purchase the large tracts of land the company owned in the state: Bladen was "a very fertile county, capable of great development"; 200,000 acres in Brunswick offered "splendid openings for capital, labor and skill"; few counties "excelled" Columbus County; and Robeson's "great advantages" made the county "very desirable to the settler."[13] Two years earlier, the North Carolina Literary Board had published a list of *Facts for Capitalists and Emigrants* to encourage the purchase of the swamplands it still had in its possession after the state legislature's 1825 grant.[14]

Yet emancipation had also turned the traditional insulation of the southern labor market from an asset into a liability. As Gavin Wright has shown, whereas the absence of a continuous influx of immigrant labor drove up the value of slave labor before 1865, after the war it made employers dependent on the free blacks they despised and turned former opponents of immigration into "avid proponents and organizers."[15] Some elite Southerners maintained that the region should be thankful for having been spared an invasion of the riff-raff that plagued northern cities. Most agreed, however, that a steady stream of "desirable immigrants" would be much preferable to dependence on black agricultural labor at a time when native white laborers increasingly sought employment in the region's emerging industries.[16] In 1870 therefore, southern entrepreneurs convened in Charleston, South Carolina to discuss plans for the promotion of immigration to the South and the establishment of information and recruitment centers. In 1884 southern leaders renewed their commitment to immigration as they gathered in Nashville, this time to launch the Southern Immigration Association.[17] In 1888 delegates from most southern states met at the Southern Inter-States Immigration Convention in Montgomery, Alabama, and again in Asheville, North Carolina two years later.[18] Several counties and states organized recruitment bureaus and sent representatives overseas to find desirable settlers and convince them to seek their fortunes in southern agriculture.[19] By the 1890s, the desire for the development of the South's natural resources *and* of the burdensome fixed capital that the cutover presented had combined with increasingly institutionalized racism to create "a case of capital seeking labor supply."[20]

Entrepreneurs and reformers alike were adept at dreaming up land settlement schemes and socioeconomic utopias in an age when the widespread availability of cheap public lands facilitated unbridled land speculation, but their implementation proved difficult, their maintenance often impossible.[21] North Carolina's southeastern corner, for example, became the prospective site for colonization plans in 1892 when "Philadelphia capitalists" planned to build a railroad from Norfolk to Charleston—through the Green Swamp—and "establish colonies of emigrants" on more than a million acres of land along the proposed line.[22] In 1894 L. C. Porter of Minneapolis purchased 15,000 acres of land in Columbus County for the establishment of a colony of Swedes.[23] Neither of these colonies materialized.

It was not until Joseph Brown, a Columbus County lumberman with a finger in a growing number of economic pies, concocted a colonization scheme for his cutover lands that an actual immigrant community came into being. Born near Whiteville in 1862, Brown moved to Chadbourn around 1880 and was one of the town's three commissioners when it was incorporated in 1883. He established a mercantile business, built a two-story hotel, and erected a sawmill, a cotton gin, and a crate factory. He was the largest stockholder in the Whiteville Lumber Company, the Chadbourn Manufacturing Company, and the Brown Lumber Company. Eventually he became the president of the Bank of Chadbourn, which opened in 1904. Business was good: "I feel like quoting from the Count of Monte Cristo," he wrote to his future wife in 1897, "when he pulled himself timid and wearily from the water . . . a free man & exclaimed 'The world is mine.'" Brown assured Minnie that this did not mean he desired always to have more, "but I feel that I can gain it"; he subsequently invested much of his profit in real estate.[24] In the early 1890s he saw strawberries growing on cutover land in Louisiana and decided, after growing them on six acres of his own land, to encourage strawberry cultivation on the cutover lands around Chadbourn. As Chadbourn's strawberry crop would be ready for market during the six weeks between the end of the Florida growing season and the end of the season further north on the Atlantic coast, its strawberry farmers would have the northeastern markets virtually to themselves for a month and a half. Brown then sought reliable farmers from outside the region to settle on lands he owned outside of Chadbourn and take advantage of this potential monopoly. He advertised in Chicago's *Farm, Field and Fireside* magazine and with the publishers organized train excursions from the Midwest to Chadbourn's soon-to-be "Sunny South Colony."[25]

Establishing as farmers native "immigrants" from New Mexico, Illinois, Iowa, Nebraska, and Michigan, as well as settlers from England and Germany who arrived in Chadbourn by way of the Midwest, Brown shipped his first crop of strawberries to market in 1896 using "pony refrigerators," galvanized ice pans that sat on top of containers to keep the berries fresh.[26] In 1897 six hundred crates of strawberries left Chadbourn for northern markets to sell for $5,562, which translated into a sizeable enough profit for such a small crop to turn most of the farmers around Chadbourn on to strawberry cultivation: the county's strawberry acreage quintupled between 1900 and 1930. The yearly shipment from Chadbourn increased from 6,000 crates in 1898 to 317,000 crates in 1906, when in one day four engines with forty-five cars of strawberries left town: a mile and a half long "red ribbon of . . . 'food fit for the gods.'"[27] In 1910 Columbus County's 2,548 acres of strawberry fields, almost 3.5 percent of the county's improved acreage, produced a stunning 4.8 million quarts.

The strawberry industry's growth rate quickly caught the attention of boosters and farmers in the vicinity of Columbus County and turned the possibilities of truck farming into a matter of public interest. The *Southport Herald* deemed the county of Brunswick "unsurpassed by any in the state" when it came to "trucking purposes." A railroad through the county, wrote the editor, would enable farmers to replace their corn crop, which only served to fatten the livestock they drove to market for an inconvenient and unprofitable twenty to forty miles, with "strawberries, lettuce, melons . . . and early vegetables of nearly every description."[28] Robeson County, too, caught the strawberry bug; several of its citizens met with "industrial agents of the Seaboard Air Line . . . to take steps for the inauguration of the trucking industry on a large scale." One week later the *Lumberton Argus* reported that the Seaboard Air Line proposed to pay for half of all strawberry plants Robeson farmers wished to plant, which had the "Raft Swamp and Moss Neck people . . . going into the strawberry business on a large scale."[29] At least according to these glowing reports and hopeful editorials, the southeastern corner's cutover was ripe for a fruitful transformation.

At the same time, the small number of immigrants who came to Brown's "Sunny South" and the relatively marginal amount of acreage that farmers around the Green Swamp allocated toward strawberry production indicate that it took more than booster rhetoric to turn the entire wasteland into a blooming garden.[30] A far more significant outcome of the search for a solution to the combined challenges of the growing cutover and the postwar labor question was a holistic vision of agricultural reform and an emerging class of entrepreneurs who sought to implement it on the region's denuded landscape as well as on its population. Brown made no bones about proclaiming that the entrepreneurial vision and organization behind Chadbourn's success were largely his: "I stood at the Depot this P. M. & surveyed the scene & . . . I felt great satisfaction at the result of three years labor. At that time this was a sleepy x roads & unknown—today the place presented a more business like appearance than any town on this R. R. with Doz[ens] of teams & hundreds of people all on the move, a town full of strangers from all parts of the U.S."[31] And among all this hustling and bustling, "my word is law: I have some influence in politics, in business, in social affairs—I usually get in a position to dictate to my business associates."[32] He frequently referred to the western settlers who came to Chadbourn as "suckers" or "deluded suckers," spoke authoritatively about land and agriculture to "expert horticulturists & fruit growers" in Wilmington, and had a hand in decisions that affected the region's development. Although one of his detractors accused him of "colossal selfishness,"[33] Brown saw things differently—he took his successes to be rewards for his struggles to overcome large obstacles against unfavorable odds. When in the midst of an election campaign he wrote to Minnie on letterhead belonging to Whiteville attorney D. J. Lewis, he used the space normally reserved for case names to write "J. A. Brown vs. Angus Wilton McLean & *the World.*"[34] It was this vision of a socioeconomic revolution led by enterprising men that inspired George F. Lucas, Superintendent of Health in Pender County, to praise Brown and express his "desire to be with you in the great work which you have mapped out for the benefit of the farmers of South Eastern North Carolina . . . The possibilities *under proper management* seem to me to be vast."[35]

"Humanics"

The emergence at the turn of the century of Hugh MacRae as the leading advocate of immigrant farmer communities reflects the growing conviction among businessmen and reformers that improvement of southern life by way of cutover colonization schemes could only be realized through farther-reaching social transformation and heavier-handed management. MacRae, a general entrepreneur from Wilmington and descendant of the MacRaes who had played such a prominent role in building the railroad network that emanated from the city, was involved in the many different projects he and his family considered essential to the economic development of the Lower Cape Fear. Concerned as he was, however, with what he considered the social, cultural, and economic backwardness that pervaded the South, he did not stop at mere business developments. During his lifetime, he joined and often headed various state and regional organizations that sought to lessen the more destructive effects of capitalist development while helping worthy southerners take advantage of its benefits. The North Carolina Forestry Association, the Associated Committees on Southern Rural Development, the Southern States Associated Committees on Reclamation, the Southeastern Council, and the Southern Immigration Association were all organizations that shared MacRae's dedication to sustained resource utilization and the improvement of southern life in the face of rapid economic growth.[36] Using these organizations as a forum in which to voice his concerns and present his economic uplift projects, he hoped to serve as a model for other southern entrepreneurs who cared about their region's well-being.[37]

Inspired by the success of Brown's Chadbourn colony, MacRae sought to tackle conditions in the southern countryside, which had gone through cycles of deterioration and improvement after the Civil War but which, to entrepreneurial southerners like MacRae, always looked bad compared to the rural North and West. "We should emphasize and hold fast to the fact that a nation is built on its agriculture," he once told an audience, expressing a sentiment very similar to that of the *Southport Standard*.[38] While industrial development had "saved the South from financial disaster," he argued, farmers who had not abandoned agriculture in favor of industrial employment had not benefited from these developments—the southern states had "the poorest agricultural system of any civilized country."[39] The quality of the South's rural life had certainly gone downhill at a steady pace since the end of the Civil War, and calls for rural reform had become louder and more numerous as the nineteenth century drew to a close. During the growth-crazed years that followed the demise of the Old South, many southern farmers, partly by choice and partly forced by socioeconomic conditions, had geared their operations toward the maximization of cash crop production for an increasingly global market. The vicious cycles of production increases and subsequent price drops that followed this decision—which in turn drove increases in farm tenancy, sharecropping, merchant monopolies, the crop lien system, and soil erosion—led to the kind of unbalanced economic growth and widespread poverty that left few aspects of life untouched.[40] MacRae, like so many Progressive reformers seeking to eradicate the outgrowths of rural poverty they mistook for its roots, partly blamed migration, poor nutrition and sanitation, substandard education, and backward medical care for the degraded condition of southern farming.[41]

Therefore, while he believed that the vastly diminished profitability of southern agriculture lay at the heart of the rural crisis, MacRae thought that agricultural reform needed to encompass both economic and social change in order to restore farming's former vitality. The farming culture with which he wanted to replace "the present obsolete one-crop system" went far beyond the mere growing and marketing of crops.[42] It encompassed intensive, diversified truck farming, under expert supervision, by farmers newly imported from Europe and the Midwest, on individual small farms located in close proximity to each other and connected through community cooperatives. This structure was to rescue southern agriculture from its increasingly unprofitable dependence on black labor and the erosive staple crop production that had sunk the region's ignorant farmers into poverty and tenancy, and also save the new breed of farmer from the isolation and drudgery that further exacerbated the old guard's degeneracy.

To make his case, MacRae joined forces with Carl Van Leuven and the Carolina Trucking Development Company (CTDC) in order to establish organized farming colonies on cutover land in the vicinity of Wilmington and thus implement a holistic vision of social improvement through far-reaching agricultural reform and social engineering. The company purchased large tracts of land along the Atlantic Coast Line's divisions in New Hanover, Pender, Brunswick, and Columbus counties, divided them into ten- and twenty-acre plots for settlers, and laid out plans for seven model colonies in the vicinity of Wilmington: Marathon, Castle Haynes, St. Helena, and Van Eden along the old Wilmington and Weldon Railroad through New Hanover and Pender counties; Maraco along the former Wilmington, Columbus and Augusta Railroad in Brunswick County; and New Berlin and Artesia along the same railroad in Columbus County. For $30 per acre, settlers interested in growing truck crops for eastern urban markets could obtain ten acres of cutover land "covered with a moderate growth of small yellow pine timber." In addition to land, MacRae told the North Carolina Society of New York, the company sold settlers houses and mules, furnished "those who are without means with seeds and fertilizers," and gave them "instructions as to planting and marketing."[43] Before their first harvest, and whenever they were in need of financial relief thereafter, settlers could turn to the company for employment in its logging and ditching operations, making use of a reciprocal relationship that accelerated the settlers' acculturation and the company's projected landscape transformation. Through such organized cultivation of the southeastern corner's cutover lands, the CTDC hoped to offer its own comprehensive yet affordable solution to the South's agricultural malaise. "A ten acre vegetable, fruit and poultry farm in the famous North Carolina Trucking Belt means a home and a business [that] insures a good living, all in one," the CTDC promised prospective settlers.[44]

The company's promotional literature and the implementation of its plans reveal the extent to which MacRae's vision was a Progressive one—he wanted to place science, diversification, efficiency, legibility, organization, and integration at the heart of the modern farming enterprise. Through the sale of small, affordable farms, the CTDC sought to instigate a shift in southern agriculture from large, one-crop farms worked by tenants to small, mixed farms worked by their owners. As MacRae depicted the problem in 1929, "Large acreage per farm . . . makes impossible either cultivation by the owner and his family or careful supervision. Most tenants or farm laborers, without adequate supervision, can not practice scientific agriculture or soil conservation."[45] The company was wary of prospective buyers who expressed interest in purchasing more than a single farm: they might initiate a system of tenancy on the company's lands or engage in the kind of land speculation that had run rampant in the South a generation earlier, facilitating the exploitative land use practices that had contributed to the skewed development of the southern economy in the first place. As

Van Leuven informed S. W. Pierce of Pierce & Company, who sought to purchase an Artesia lot to add to the tract he already owned, "it has been our policy . . . to sell to one person only a limited area so that . . . they would actually develop the land into a farm . . . our plan contemplates more than simply obtaining money from the sale of our land."[46]

The company also envisioned a shift from old-fashioned, wasteful farming methods to scientific, efficient agriculture. The colony format was the kind of orderly communal project with which many Progressives sought to curb the ruthless individualism and selfish greed that had characterized the Gilded Age. First, the company granted settlers easy payment plans for their lands—one-fourth in cash, the balance paid in three yearly installments without interest if paid when due—and helped them obtain the necessary implements at affordable rates. This alleviated some of the initial hardships of establishing a farm that had gotten many local farmers in debt to powerful merchants and landowners. Second, the different planting times and harvesting seasons of truck crops increased efficiency as diversified farms guaranteed year-round employment and eliminated the winter lull that might drive farmers to unproductive behavior. Farmers who grew beets, lettuce, turnips, spinach, onions, cabbage, eggplant, cucumbers, okra, peas, beans, squash, radishes, tomatoes, sweet potatoes, Irish potatoes, melons, and strawberries did not have to fear total failure since their risk and their income were spread over various crops and harvest periods. Third, cooperative efforts served to restrain unhealthy competition within the colonies, cut costs, strengthen the farmers' marketing positions, and create a local socioeconomic safety net so farmers would not fall into the poverty and depravation that afflicted so many southerners. The stipulation that only men with families could settle in the colonies—"single men are apt to become dissatisfied," explained *The World's Work*—underscored the need for strong social and communal ties within the colony. Fourth, the company worked to ensure that farmers employ proper farming and marketing methods, using both education and supervision to guide farmers to prosperity. It maintained a demonstration farm and employed a cadre of instructors to advise settlers in matters of planting, cultivating, and harvesting crops. It entrusted the day-to-day supervision of each colony to a superintendent, a "business adviser" who kept the settlers abreast of market conditions and instructed them in such matters as packaging and shipping.[47]

In addition, Progressive faith in the value of scientific expertise guided the entire transformative process from the selection of cutover sites to the marketing of the crops. As he searched for usable land, MacRae had engineers prepare soil reports and maps of available tracts in order to determine their location, size, existing improvements, drainage, overall health and, most important, soil types and possible crops.[48] And rather than limiting the use of scientific data to the preparatory stage of the colonization project and risking a relapse into traditional southern farming methods once settlers commenced cultivation, MacRae encouraged colonists to capitalize on scientific findings of this nature. The company propagated limited farm size, not only to ensure that farmers would be able to attend to all aspects of their enterprise themselves, but also to enable the implementation and company supervision of scientific farming methods. Indicative of this aspect of the project was MacRae's advocacy for the adoption of modern pig farming, away from traditional open range feeding and corn supplements, and toward pasture feed crops such as soy beans, peanuts, and bur clover, as well as toward scientifically derived diets for the different pasture seasons. By following specific feeding guidelines, farmers could fatten their hogs at a predicted rate and "have money coming in twice a year."[49]

Proclaiming a positivist vision of a society in which all aspects of life were integrated within a legible, corporate whole, the company further emphasized the quality of local health conditions, roads, and schools, and assured prospective settlers that the region had the infrastructure necessary for the successful integration of the farming enterprise into the larger regional economy. With all its property located on the Atlantic Coast Line Railway, shipping facilities were excellent, the company wrote, especially since the railroads devoted "special attention" to the handling of the very profitable traffic in truck crops. With Wilmington nearby, settlers would have a ready local market for their crops and for their poultry and eggs. Wilmington was also an important node in the transportation network that moved truck crops from the southeastern coast to the insatiable markets of New York, Philadelphia, Baltimore, and Washington, D.C., where demand for out-of-season fresh vegetables was ever on the rise. Moreover, the port city offered good export facilities to those who wished to grow cotton in addition to their truck crops, as well as five banks to provide farmers with financial assistance. Overall, the company presented Wilmington as the kind of city that offered the farmer those amenities of modern urban life that, when available at the proper distance, improved the quality of life in the country.

The most radically "progressive" aspect of MacRae's effort to lift southern agriculture out of its rut was the recruitment of Midwestern and European whites for settlement in ethnically homogeneous farming colonies to bring about the racial uplift of the white farming population and gradual displacement of the black labor force. The CTDC printed copies of *Wealth in Carolina Truck Farms* in Dutch, Danish, Hungarian, Italian, French, German, and Slovak and distributed them in Europe and the United States to attract immigrants. In his address before the North Carolina Society of New York, MacRae asserted that "there is no quicker way to increase the wealth of a country than to add to its industrious and thrifty population." Proof of this abounded in the northern and western states, where millions of "sturdy immigrants" from all over Europe had stimulated agricultural, industrial, and urban development. The South had not benefited from this influx of labor because the surplus of black labor, the low wages and purchasing power this generated, and the resulting weakness of industrial development steered immigrants to regions that offered better opportunities. MacRae said that the problem was such that steamship and railroad agents designated the South as "the Country of the Black Man"; he claimed that as late as 1908 immigrants still feared cannibalism in the South.

Emblematic of the notion among white entrepreneurs that the black population constituted an obstacle to economic development was one Southerner's claim that "the elimination of the Negro from politics makes it possible to consider governmental problem from an *economic* standpoint, and the conditions now . . . are just the results that were sought . . . by the revolution of '98."[50] Hugh MacRae had himself been one of nine Wilmington businessmen, the Secret Nine, who had helped foment racial tension in the city during North Carolina's white supremacy campaign in 1898, specifically calling for the reclamation of *labor* as a realm of white dominance. On 10 November of that year he had helped lead a white mob on a murderous path of violence through Brooklyn, Wilmington's black neighborhood, allegedly boasting of ninety African-American deaths in its wake. But because disfranchisement and violence proved insufficient for structural economic change, MacRae hoped that the settlement of an advance guard of immigrants would stimulate further white migration on such a scale that the newcomers would solve "the race problem of the South" or minimize it to such a degree "that its solution will not be a matter of supreme importance."[51]

The envisioned establishment of immigrants as farmers to replace whites who had abandoned agriculture for employment in mill and town was an obvious departure from the migratory patterns of the immigrants who fueled the growth of the North's emerging urban-industrial centers, as well as a blatant attempt to squeeze African-Americans out of agricultural and industrial labor. MacRae figured that if industrial workers were loath to come to the South because wages remained low, the answer was to draw desirable immigrants such as northern Italians and Dutchmen to southern agriculture with promises of landownership and prosperity. The "supply of better labor" that these immigrants constituted, according to the *Manufacturers Record*, would "make the negro incidental instead of the controlling factor" in the southern labor market. The paper contrasted pictures of straight rows of lettuce in front of equally neat immigrant homes with a photograph of the rickety shack occupied by a black family and described the latter as "the kind of dwelling that must disappear with immigration of more whites to parts of the South."[52]

In order to protect the racial integrity of the colonies, the CTDC prohibited the sale of land to blacks[53] and the company did not stop there: southern whites initially found the colonies closed to them as well. MacRae also sought to maintain ethnic homogeneity in the colonies to ward off the potentially bad influence of traditional southern farming methods, use the cultural unity of a close-knit, all-immigrant settlement to alleviate the traditional boredom and loneliness of farming, and smooth the settlers' transition from Old World to New. As interest in his projects grew, he decided that what became the Marathon colony should be exclusively for Greek immigrants, Castle Haynes for Dutchmen and Hungarians, Van Eden for Dutch settlers, St. Helena for Italians, Artesia for Englishmen, and New Berlin for Germans.[54] Separation from native farmers and close supervision by a company superintendent, therefore, not only facilitated the implementation of modern farming methods; the immigrants' shared ethnic backgrounds also served to stimulate a cooperative spirit within the colony, the kind of socialization the company further encouraged through aid for the erection of community halls, churches, and schools. This type of Progressive, ethno-social engineering had already produced settlement houses in the North and Indian boarding schools in the West;

but in the South the organization of such improvement initiatives had not to any large degree accompanied the political disfranchisement and economic subordination of blacks. Yet Hugh MacRae felt that it was exactly the South that stood to benefit most from what he called "humanics," or human engineering. "The men who will study our complex, often deformed mass-life," he proclaimed, "and will reconstruct it on a basis of mutual adaptation between all units . . . will be Human Engineers of a new society."[55]

"An empire of devastated areas into prosperous ranches and homes"

By the time Hugh MacRae was helping immigrants get settled in their neat little houses, cutover cultivation, agricultural reform, and communal engineering had taken on regional, national, and international dimensions. In much of the western world increased recognition of the close connections between, on the one hand, commercial development of natural resources and, on the other, the growth and development of societies whose economy no longer "simply" revolved around agriculture or extractive industries, boosted efforts to integrate and increase the productivity of land and people amid booming manufacturing and trading centers. Advocates of such efforts tended to cast their analyses and plans in progressive terms of corporate uplift, science-driven improvement, and social change. The governments of Australia and New Zealand engaged in experiments to market products and attract capital, aiding farmers in the expansion of their crop base to include wheat, meat, and dairy. In New Zealand, milk became a symbol of a green environment in which to raise families.[56] Among similar projects in the Netherlands was Walden, the fifteen-acre commune Dutch author Frederik van Eeden had established thirteen miles from Amsterdam. After its failure Van Eeden came to the United States to found a "Cooperative Company of America," joining forces with MacRae in eastern North Carolina's "Dutchman's paradise" to create the Van Eeden colony, which eventually became a refuge for Jewish escapees from Nazi persecution.[57] In Brazil, an elite of Portuguese descent recruited white Europeans to replace the labor of black slaves after emancipation in 1888 and so "whiten" the population, while Italian immigrants in Argentina made up the bulk of the wheat farmers who turned the pampas into the world's bread basket.[58] So widespread was the desire for rural reform that the manager of Labor Information for Italians in New York City had the honor to introduce Hugh MacRae to Italian Countess Gigi Ciprichi, who was "deeply interested in the colonization problem."[59]

In the United States proponents of settlement around the Great Lakes had defeated reforestation advocates in a heated debate over the future of the white pine cutover there.[60] In the state of Washington, entrepreneurs developed plans to establish farming communities after efforts to restore the western woods to their former glory had run into a wall of indifference.[61] While advocates of rural reform were either "urban agrarians" who sought to heal the ills of urban-industrial America by moving suffering urban dwellers to a more healthful life in the country while keeping country folk from leaving, or "anti-agrarians" who saw farmers "as the embodiment of all that was crude, selfish, ignorant, irresponsible and uncontrolled" and sought to improve rural life in order to lift farmers out of backwardness into the modern age, neither party propagated a return to wild nature. The reformers of the era's "back-to-the-land" and "country life" movements shared a wholehearted commitment to modernization and development as the way forward, a desire to modernize farm life and make it more attractive through an infusion of the blessings of urban-industrial society. They strove to alleviate the hardships that had traditionally attended wresting a living from the soil and replace them with a readily available network of local and regional economic ties that typically emanated from the city. Using the technology of transportation, surveys, education, business efficiency, and marketing, agricultural reformers thus sought to help farmers help themselves and build a healthy, prosperous, and modern countryside.[62]

Indicative of the perceived interconnectedness of modern science and technology, landscape transformation, rural health, and agricultural prosperity were renewed efforts to rid the country of its most drab, sickly, and unproductive environments: the swamps and marshes that supposedly "affected" city and countryside alike. State governments across the nation committed resources to turn such diverse environments as the Florida Everglades, the prairie potholes of the Midwest, and California's Central Valley into productive farm regions at the expense of the areas' rich biodiversity.[63] North Carolina tried to salvage its own swampland project after the State Board of Education lost claim to its largest swampland areas. The Supreme Court, in *The Eastern Land and Lumber Company* v. *the State Board of Education* (1887), had declared invalid the way in which much of the land had come into the state's possession—through tax deeds by sheriffs to the governor. Engineers kept setting up camp to work at drainage projects, but the board increasingly struggled to claim land for drainage and sale purposes.[64] The piecemeal drainage of wetlands did little to realize North Carolina Progressives' original dream of converting "loathsome marshes and dismals" into "fruitful fields" and "the delightful habitations of man." In response, Hannibal Godwin, one of the state's representatives in the U.S. House from 1907 until 1921, called for federal involvement in swampland drainage. He told the House that the existence of large areas of swamp and overflowed lands constituted not simply a problem in terms of unused resources but also "a menace to human prosperity, health, and life."[65]

To the age-old problem of unproductive swamps was now added the uselessness of the growing cutover. Southern lumbermen, who as owners of much of the cutover intimately knew its strain on economic development, were among the most prominent advocates of actively stimulating the agricultural development of the lands they had denuded. Calls for reform from lumber interests, too, tended to emphasize the totality and diversity of the farming enterprise and the reform of traditional farming methods. The *Southern Lumberman*, one of the industry's main mouthpieces and a supporter of selective immigration, became a forum for discussions about the possible transformation of the cutover through cultivation, population, and a corporatist vision of a legible mixed economy. James Kirby, lumber magnate and the journal's vice president, suggested "stimulating the agricultural development of these lands and . . . attracting an influx of domestic and foreign immigration skilled in agricultural pursuits and financially independent."[66] The journal held up the example of Lockhart, Alabama, where W. S. Harlan of the Jackson Lumber Company established forty-acre crop farms on 5,000 acres of cutover land. The project also included a pecan tree nursery, a poultry farm, and a horse farm, while about 1,600 head of cattle grazed an additional 21,000 acres of cutover pasture. The cattle supplied the town and the logging camps with beef that formerly had to be imported from the West, "keeping the money at home." The management also bred hogs and encouraged local farmers "to raise a better breed of hogs and to do away with the time-honored 'razor back.'" In fact, the *Lumberman* reported, all over the South lumber and railroad companies encouraged settlers to take up the cultivation of such non-traditional crops as cucumbers, tomatoes, and pecans, as well as the raising of poultry, beef and dairy cattle, goats, and sheep.[67]

The Southern Pine Association, which tended to be critical of the "cut and get out" mentality of sawmills and instead promoted a progressively organized and self-regulating industry whose methods would stimulate integrated economic development, had from its inception also maintained a committee whose primary concern was the utilization of the cutover.[68] What could lumbermen do, O. H. L. Wernicke asked on the association's behalf, to fulfill what he considered their "obligation to the South" and make their cutover lands productive again? The proposed colonization of soldiers and sailors fighting in World War I at the time of his writing was only an idea, and even if attempted would not begin to solve the problem. The processing of the stumps, "fat wood," limbs, and small timber strewn about the cutover mostly failed to turn even a small profit. And in any case, the quality of much of the cutover soil was insufficient to reward the efforts of any small farmer.[69] Instead, what lumbermen could and should do was establish the cutover as pasturage for livestock, which would allow the South to "follow in the footsteps of more developed sections of the country which first went through a period of livestock raising on the open range before intensive farming ensued." Grazing, Wernicke contended, improved the quality of the soil, encouraged the growth of grasses, made stump removal easier, and reduced the occurrence of fire. In his grand vision of the transition from forest to farm, livestock would convert "an empire of devastated areas into prosperous ranches and homes." And no animal was better suited to bring about the cutover's remarkable transformation than the sheep, the "woolly little friend of the human family." Its needs were minimal, its habits good for the land, and its products—mutton, wool, tallow, pelts, and bone—always in high demand. In Wernicke's view, sheep were "almost like money in the bank."[70]

As adherents to Progressive ideals of order, efficiency, economic integration, and social uplift gained ascendancy in local, state, and federal governments, agricultural reform became a focus of policy debates in legislative sessions, committee meetings, and conference panels across the United States. In North Carolina Wade H. Phillips, director of the Department of Conservation and Development, commented on the multiplicity of the state's agricultural problems and stressed the need for change in all aspects of rural life: "Our landowners have had a hard time recovering from the old regime of negroes and cotton," he wrote; "many of them are still moving in the old grooves." He suggested that farmers diversify their crops so that their lands might remain healthy and productive. He recommended stimulating animal husbandry as farmers did not own sufficient numbers of milk cattle, pigs, and fowls to sustain a healthy domestic livestock industry. He also felt that if farm life were more attractive, young men would no longer leave for the city in numbers large enough to bring about a shortage of reliable laborers and future farm owners. Improvements in terms of roads, schools, and electricity had had some effect in this regard, and Phillips expected the development of electrical machinery to take some of the heavy work off the farmer's hands. Still, he envisioned the establishment of a number of model farms where "a group of thrifty farmers from the northwest or even from Europe" could show North Carolinians how best to implement modern, scientific farm techniques in a less individualistic, more communal setting.[71] His concerns were echoed by C. B. Williams, head of the Department of Agronomy at the North Carolina State College of Agriculture and Engineering, and by B. W. Wells, head of the school's botany department, who argued for agricultural and rural reforms along similar lines.[72]

Eventually the U.S. government, which had staged scientific farming demonstrations through the Department of Agriculture since 1898, responded to calls for cutover development and wetland drainage in the South.[73] It shifted its attention from gathering and conveying information through agencies like the Bureau of Drainage Investigations to the attempted creation of farming communities through the Bureau of Reclamation. Established in 1902 under the Newlands Reclamation Act, the bureau played a crucial role in water management in the West, building dams and other irrigation projects in seventeen western states. Under the leadership of Elwood Mead, who became Commissioner of Reclamation in 1924, the bureau also began to look southward. While reformers like MacRae kept up their chorus of disenchantment with the vast expanse of southern wastelands, Congress in late 1924 appropriated $100,000 for a survey on the possible development of swamp and cutover timberlands, and Mead seized the moment to investigate reclamation possibilities in the South. While serving as professor of Rural Institutions at Berkeley and as chairman of the California Land Settlement Board between 1915 and 1924, Mead had headed the State Demonstration Colony at Durham—a model farm colony the state legislature hoped would help improve rural life in California.[74] He greeted with enthusiasm, therefore, suggestions to establish under federal auspices similar colonies on drained cutover lands and help southern men "become farm owners in organized communities."[75] Mead considered the greatest problem of southern farming a lack of landownership and organization among the region's farmers: "where the land is cultivated, it is being farmed by hired negro laborers, or by tenants lacking capital, skill or ambition." What southern farmers needed were "modern scientific methods in agriculture, cooperation in business and social affairs, and a system of credit which will give to men of small means the opportunity to buy a home and have a lifetime in which to pay for it." With a possible $10 million in funding provided for in a proposed reclamation bill, the role of the federal government would be to purchase 5,000 to 10,000 acres of suitable land in each southern state, prepare them for occupation and cultivation, and divide them into small farms. Since the government had helped just "about everybody else," Mead and his supporters figured it was time the farmer received his share.[76]

The prospect of the federal government investigating and possibly buying swamp and cutover lands in turn sparked the interest of private landowners eager to shed unproductive acreage as well as of state government officials eager to see federally stimulated economic development. F. L. Finkenstaedt, president of both the Waccamaw Lumber Company and the North Carolina Land Owners' Association, was among the many lumbermen who saw in every state and federal settlement plan an opportunity to shed their heavy burden of cutover land.[77] All tried to persuade the Bureau of Reclamation of the suitability of certain sites with an appeal to the corporate vision of socioeconomic reform that had, by the 1920s, become de rigueur. The Butters Lumber Company was among those that filled out a questionnaire as to the character of its land in order to alert the bureau to the perfect colonization site. The company described 15,000 acres in Columbus and Robeson counties that would, with the proper drainage, support grazing and grow the "usual crops"—corn, tobacco, cotton, oats, and potatoes. These crops would find ready markets in Chadbourn, Bladenboro, Fairmont and Lumberton, all within fifteen miles of the property. Having been the property of a lumber company, the land had several other advantages. The Conway-Elrod branch line of the Atlantic Coast Line ran through the property, and Boardman, a logging town, already had fifty "white dwelling houses" and seventy-five "negro dwelling houses" as well as a large store, a post office, warehouses, and a railroad station. In addition, President John Merriam explained to Elwood Mead, second growth shortleaf pine covered some of the land, increasing its value.[78]

Wade Phillips of the Department of Conservation and Development also sought to include North Carolina lands in the Bureau's reclamation plans and asked C. B. Williams to go over the available information on tracts in Pender, Hyde, Robeson and Columbus counties. Given the dearth of information on the latter two, Williams sent a representative, W. B. Cobb, who led him to the conclusion that the cutover lands of the Pate estate in Robeson County were "best suited for the purposes of the Bureau of Reclamation." Like the Butters Lumber Company, Cobb listed soil types, crop possibilities, easy drainage and clearing, the absence of extensive swamp- and wasteland, the proximity of transportation facilities, markets, and banks, and opportunities to establish cooperative buying and selling organizations as indications that the tract was the perfect site for a progressive farming settlement project.[79] And when proponents of agricultural reform gathered in Washington, D.C., to attend reclamation conferences and rally around the idea of federal involvement in cutover development, they revealed the extent to which attempts to bring about socioeconomic changes in farming societies were an international phenomenon. The delegates, among them North Carolinians Hugh MacRae, Wade Phillips, Joseph Brown, and G. A. Cardwell of the Atlantic Coast Line, pointed to the success of the various reform projects in Denmark, the Netherlands, Canada, South Africa, Australia, and New Zealand.[80]

Between 1890 and 1930, then, North Carolina's southeastern corner was touched by a global reform movement whose principal advocates sought to channel the large flows of human migration that were changing social landscapes everywhere and direct them to areas plagued by illegible and unproductive farming practices and labor relations. Local and regional boosters were quick to see in every settler from outside the area a harbinger of the migratory wave and agricultural revolution to come. "The efforts of the Carolina Trucking Development Co. . . . are meeting with great success," cheered the *Wilmington Messenger* on 12 March 1908. Besides two families who settled in Castle Haynes and an Italian family of six who settled in St. Helena that day, "four German families consisting of 14 people . . . will go out on the afternoon train to New Berlin." The paper described the immigrants as "a very high class of people, far above the average immigrant" and congratulated Wilmington for "securing such people to help develop her back country."[81] That same year *The World's Work* proclaimed that "Mr. MacRae has opened a new door of hope"—hope that the southern states would "become rich with an increasing speed."[82] The *Manufacturers Record* ascribed to MacRae near-messianic qualities, rejoicing that he "raise[d] men from a condition of practical serfdom, of misery, and of poverty without hope, to one of independence, of happiness and prosperity."[83] Lumbermen, newspaper editors, land developers, and anyone with an interest in the economic development of the Green Swamp region along capitalist lines kept up a chorus of expectation that a vastly improved rural society was just around the corner.

"I regard all my settlers as my children"

Yet the expectant voices of praise faded quickly around the Green Swamp as efforts to emulate the success of Joseph Brown's Chadbourn colony almost all came to naught. MacRae's Maraco project, in the planning in 1908, never came off the ground.[84] The Artesia colony did not outdo Maraco by much: by 1920, the Dutch entrepreneurs who had initially supported the settlement had abandoned it.[85] Only four of the village's sixteen households, or twelve out of eighty-nine people, originated outside the region: an Austrian-born family of three, a Hungarian couple, a couple from Illinois, and a Hungarian family of five. It appears that MacRae had managed to sell only ten individual farms to immigrants who already resided in the United States, out of a total of 133 lots he intended to sell to settlers from abroad. Most of the immigrants moved away, and Artesia eventually ceased to exist altogether.[86] New Berlin fared a little better—renamed Delco during World War I to avert anti-German hostility, it survived as an independent town. By 1920 the town boasted a sawmill with a foreman, two laborers, and an inspector, and served as a shipping point for turpentine and resin. According to a report in 1937, by which time Hugh MacRae & Company had sold only about half of the town's property, the colony provided settlers with a train station, a telegraph and post office, a public school, several churches, and various stores. But despite the endorsement of *The World's Work*, which ranked the Germans of New Berlin "second to the Italians as immigrants," of fifty-two households in 1920 only four individual settlers were Germans. Delco was also home to a Hungarian man, a Hungarian couple whose two children had been born in Canada and Ohio, a single Dutchman, a Michigander of English descent with an Ohio-born wife and son, families from New York and Maryland, and a Polish girl; but the town was not the thriving immigrant farming community MacRae had envisioned: native North Carolinians still made up the bulk of the population.[87]

Figure 9. *Wealth in Carolina Truck Farms*, a promotional pamphlet in Italian from the Carolina Trucking Development Company. (Courtesy of Hugh MacRae Papers, Duke University.)

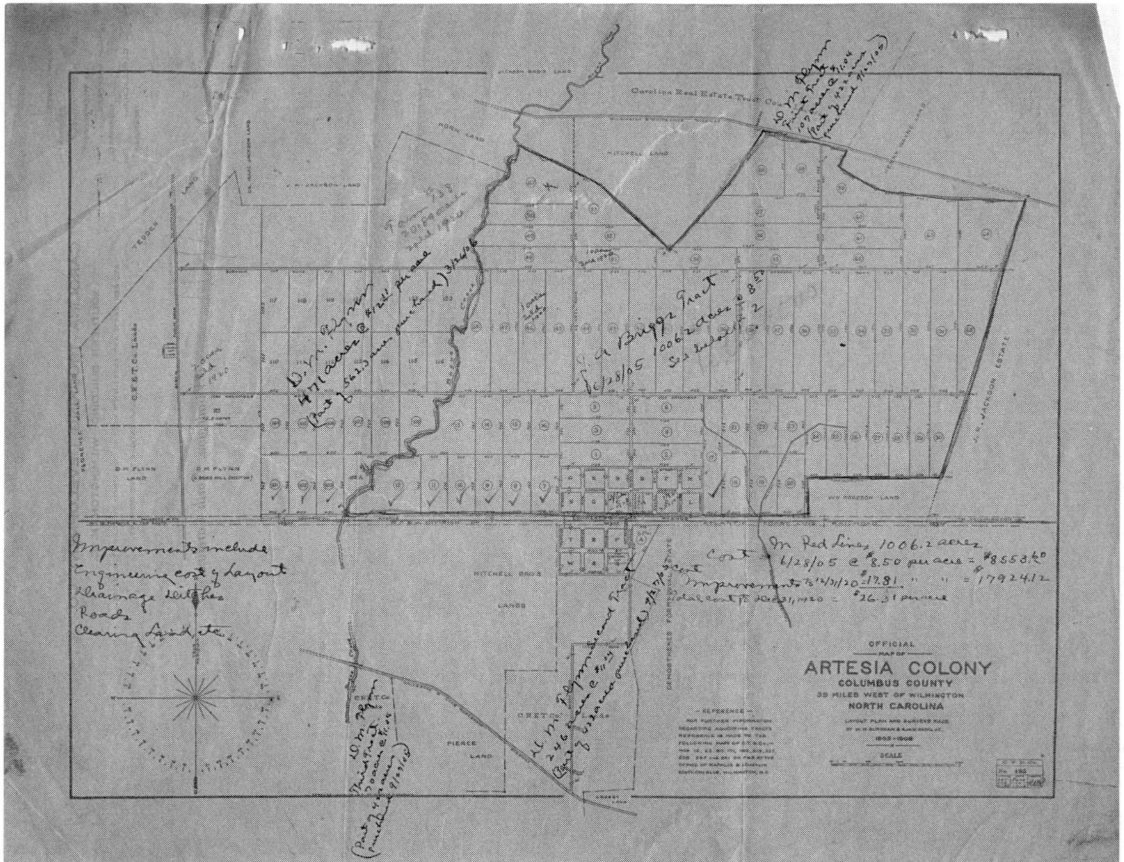

Figure 10. Plan of the Artesia Colony, 1905–1908. Note how the colony is laid out in rectangular plots along straight lines, as well as the small number of plots checked as having been sold. (Courtesy of Hugh MacRae Papers, Duke University.)

Indeed, with the exception of the strawberry boom in and around Chadbourn, colonization and reclamation efforts overall left agriculture and rural life around the Green Swamp largely untouched in terms of what their proponents sought to accomplish. By 1930 the combined number of foreign-born whites in the four counties equaled 175, hardly the stuff of racial revolutions in a region that was also home to 12,768 American Indians and 49,502 African-Americans. Although improved farm acreage increased in relative terms as total farm acreage declined, farms in 1930 still averaged 67.1 total acres, 43.4 of which remained unimproved, bearing witness to an overall failure to implement on a large scale more intensive farming methods on smaller farms. At the same time, the decrease in average farm size meant that the sheep or cattle ranch, which Wernicke and other lumbermen had envisioned as the agent of revolutionary change, also failed to rise to prominence. Livestock constituted less of a presence around the Green Swamp after the lumber boom than before—the number of sheep alone dropped from 23,218 to 5,587 between 1890 and 1930. The only animals to roam farm and field in ever-larger numbers were those mainstays of traditional southern farming, the mule and the ass. Farmers also stuck to the cultivation of traditional cash crops, apparently switching from rice to cotton to tobacco when potential profitability so directed them. By 1910 rice production had dwindled to a fraction of nineteenth-century levels while cotton production and acreage increased steadily until 1920. After 1920 both decreased in Bladen, Brunswick, and Columbus counties, while Robeson's cotton acreage was still expanding in 1930. In all four counties tobacco production and acreage, both in absolute terms and as a percentage of overall improved acreage, rose steadily through 1930. Local newspapers frequently expressed pride in the region's lively tobacco markets, reflecting the central place these staple crops had come to occupy in the region's economy.

To be sure, a shift in the region's agricultural patterns toward diversification and increased truck farming did occur, but it is unclear to what extent the efforts of the Department of Conservation and Development, the Bureau of Reclamation, and Hugh MacRae had contributed to this particular shift in the region's agriculture. Farmers in the region did increasingly plant non-traditional truck crops for the growing urban markets on the eastern coast. Expanding the acreage for fruits and garden vegetables while applying more productive and intensive farming methods enabled farmers to increase both total production and yield per acre.[88] In addition, farmers increasingly grew feed crops for their livestock as they abandoned the grazing of animals on their woodlands. By 1930, farmers used less than 10 percent of their 475,455 wooded acres as pasture. During the same period the number of horses and mules increased, but the number of cattle, hogs, and sheep on farms decreased much more dramatically: farmers were growing more food for fewer animals. Even when hay crop acreage declined between 1920 and 1930, the total number of hay-consuming horses, mules, and cattle declined to such an extent that farmers effected a 17.3 percent increase in hay acreage per animal.[89]

Such changes fell far short of the goal Elwood Mead claimed for the reclamation movement, which was "to restore to the South the kind of agriculture and rural life it enjoyed before the Civil War." But Mead and his fellow reformers too often ignored two central aspects of antebellum rural life—the importance of black slavery and the abundance of forests. If imagining a past South without a large black or Indian presence was relatively easy during the halcyon days of American Nativism, the racial cleansing that was at the core of the South's agricultural reform ideology proved much more difficult to realize. Instead, the socioeconomic dilemmas of race, which had spurred the drive for reform, continued to overshadow the discourse and practices of economic development, effectively obscuring from reformers' view the actual working and living conditions of farmers around the Green Swamp. Reformers' incessant hammering on the degenerative effects of traditional southern agriculture on rural life notwithstanding, farm tenancy around the Green Swamp increased only slowly, even among non-white farmers. As late as 1930 nearly three-quarters of all farmers in Bladen, Brunswick, and Columbus owned their farms. Among non-white farmers, who as a group experienced tenancy at a slightly higher rate than white farm operators, ownership was common as well: over two-thirds of all black farmers were full or part owners of their farm. Non-whites owned almost one-fifth of all farms in Bladen, Brunswick, and Columbus. None of these three counties experienced tenancy to the extent that it plagued the state as a whole, let alone states such as Mississippi and South Carolina—tenancy there already topped 60 percent during the 1920s.[90]

Robeson, by 1930 home to ninety-seven percent of the area's Indians and 46 percent of the area's African-Americans, was the only county to suffer higher-than-average tenancy rates, likely as a result of its racial makeup, with over 70 percent of all farms not operated by full or part owners. With a large non-white population, high tenancy rates, and a greater commitment to cash crops than the other counties around the Green Swamp, Robeson County showed all the "southern" characteristics that warranted reform. Yet it was exactly the race issue that convinced reformers *not* to establish a farming colony in Robeson County. When Cobb's report on Robeson County and Phillips' recommendation of Pembroke as a prospective site caught Elwood Mead's attention, the commissioner met with Wade Phillips and Hugh MacRae in early December of 1926 to check out the place for himself.[91] The Pembroke development plans went straight out the window, however, when local whites began to protest the possible inclusion of Pembroke's native Indian population. Pembroke was home to a large number of Robeson's 9,000 Indians, and many of them attended the Indian Normal School there. A. F. Corbin of the school's Agricultural Department saw how young Indian men who completed course work at the school tended to show real interest in progressive farming and management methods, while students whose parents took them out of school for farm work and kept all their earnings discouraged them from continuing their education. He was convinced that implementation at Pembroke of MacRae's "plans for better living at home and more cash crops" would encourage more young men to settle "among [their] own people." Linking scientific farming to prospects for improved rural living, he predicted that "authoritative expert advice," crop rotation and diversity, cattle, hogs, and the use of natural fertilizer could produce not just prosperous farmers but "a happy and contented community."[92] Mead, however, while acknowledging that Indian farmers "would make highly desirable settlers on such a project," did not share Corbin's agrarian/utopian vision for the local Indian question. He worried that focusing on Indians would open up the question of race, because locals were mostly worried about the mixed settlements that would result from Indian inclusion. Combining whites and Indians in a colony only to enforce state laws regarding intermarriage and segregation would, according to Mead, undermine the "highly essential community enterprises and community spirit." People also wondered if it was appropriate to choose an all-Indian settlement over an all-white settlement. "Even if successfully established," Mead wrote, "the effect of such a settlement upon the conditions of rural life in the State at large could not be as outstanding as that of a more representative white colony." He also claimed that finding a hundred qualified Indians was more daunting than finding an equal number of desirable whites.[93] Phillips agreed and scrambled to nominate another North Carolina tract, eventually deciding on lands near

Lake Waccamaw whose owner, the Carolina Real Estate Trust Company, had already tried to convince the federal government that its property was preferable to the Pembroke estate because the local community was mostly white.[94]

Even more crucial to the overall failure of farm reform was the traditional role forestland ownership continued to play in rural life. Patterns of landownership and improvement were indications, not of what MacRae considered "badly organized farming and a dreary rural life," but of farmers purposely holding on to more land than they could plant in crops.[95] The fact that farmers on average cultivated 23.5 acres of improved land and owned another 43.4 acres of unimproved, mostly wooded land suggests that farmers purposely refrained from maximizing crop production, opting instead for a mixed economy of crops, timber holdings, and forest labor. Despite attempts to change farming methods around the Green Swamp, one-third of the entire area in 1910 consisted of farmer-owned woodlands, and almost 68 percent of all farmland in 1930 remained unimproved. Despite the widespread opportunity and, undoubtedly, pressure to sell timberland to the lumber industry, farmers held almost 92 percent of their unimproved acreage in woodland. More than one-fifth of the land area in Bladen, Brunswick, and Columbus consisted of woodland that farmers owned and used for cutting firewood, saw logs, fence posts, railroad ties, and pulpwood, as well as for hunting and, if necessary, for the creation of crop or pasture land. While reformers held up sharecroppers as an example of the agricultural backwardness that only modern scientific farming could erase from the southern landscape, the majority of farmers in the southeastern corner clung to their woodlands as a bulwark against the poverty of monoculture.

Imagining a region where farming no longer combined livestock and crop cultivation with forest extraction under the roof of a single farm operation proved easier than convincing farmers to relinquish their woodlands. Atlantic Coast Line agent Allen Maull tried to attract farmers to southeastern North Carolina by stressing the region's abundance of sawmills and lumber, suggesting that new settlers would not have to engage in the heavy preparatory work of logging and hauling timber.[96] But farmers in the southeastern corner did not want their involvement with the lumber industry limited to mere consumption of lumber products, even if they did find their role in the forest industry confined to only a few aspects of the market process in an economy increasingly governed by rules of supply and demand, productivity, divisions of labor, and the availability and use of capital. Unable to compete with the industry's specialized technology and large-scale command over labor, a declining number of farmers augmented their operations as outright sellers of cut timber, sawed lumber, or naval stores, engaging mostly in the production of firewood for domestic use. Still, southeastern farmers continued to benefit, both on their own farms and within the regional economy, from a mixture of improved, slightly altered, and unimproved acreage that produced goods for home consumption and for regional crop markets, as well as for the naval stores and lumber trades. Robert Smith, for example, cultivated a 660-acre farm in Bladen County that supported a mixed growth of sweet gum, tupelo gum, cypress, sycamore, ash, hickory, and oak. According to a frequent visitor to his farm, Smith, like many farmers in the region, "cultivated parts of the land, and worked turpentine . . . cut timber on parts of it, [and] used parts of it for pasturing."[97]

In addition, farmers who continued to own a significant portion of available timber resources left the clearing of their timberlands to logging crews and sawmills, reaping financial gains and indeed saving what was often backbreaking labor. And as the production and marketing of forest products for commercial purposes shifted largely from farm to distillery and mill, growing numbers of farmers engaged in forest exploitation as wage laborers rather than as independent producers.[98] As long as enough forest remained to sustain recreational and subsistence practices as well as a lumber industry that hired farmers for occasional labor or supplied them with the lumber they had previously produced themselves, farmers stood to benefit from cleared lands and the opportunities the growing industry offered. When reformers sought to diminish farm size and increase productivity per acre, therefore, they misidentified as the ubiquitous inefficiency of southern farming a central aspect of farm operations and farm life in forested areas.

If the Green Swamp region did experience an agricultural crisis, it was one precipitated by a decline in the overall wooded farm acreage that sustained the region's particular economic organization, especially in terms of the access tenants and sharecroppers had to privately owned timber resources. While farm ownership remained high, and production, productivity per acre, and farm values increased (all signs of the relative overall health of agriculture), the amount of wooded farmland farmers could harvest for timber or turn into pasture decreased by nearly a third between 1880 and 1910 and again by nearly half between 1910 and 1930.[99] The amount of unimproved acreage owned by farm owners fell by more than a third between 1910 and 1930, from ninety-two to a little under fifty-seven acres. Tenants and sharecroppers had, on average, access to one-fourth the amount of unimproved acreage farm owners possessed; by 1930 this meant an average of fourteen acres. In Robeson County, tenants cultivated on average the same amount of improved acreage as farm owners; whereas the latter owned an already comparatively low average of forty acres of unimproved land or nearly thirty-five acres of woodland in 1930, the former had access to only one-fifth that amount. At least in terms of improved acreage, increased tenancy translated into a higher degree of crop cultivation; however, in an area where access to forest resources was an essential aspect of rural life, tenancy tore at the fabric of life more than the alleged non-productivity of too much farmland. Much more than woodland-owning farmers, tenants and sharecroppers felt directly the restrictions placed on farmers by the very industry that was supposed to be a model for opportunity and progress. Because they did not own the land they worked, they could not sell the timber on it. Because they had access only to limited woodland acreage, they increasingly depended on the local market for their firewood and lumber needs. And as a result of the consolidation of private woodland in the hands of non-farmers, their access to the forest's supplementary sustenance resources also declined.

The tension between reformers' visions of agricultural progress and the persistence of more traditional farming practices, then, was really a tension between different visions and practices of community, one a scientifically engineered and supervised, ethno-social economic community divorced from traditional forest use, the other a more traditional rural community of which the cultivated land, the forest, and the swamp were all indispensable parts. To be sure, when L. H. Duncan, secretary of the Southeastern Council (of which Hugh MacRae was also an active member), enumerated the various concerns that shaped his attempts at improving agriculture around Wilmington, he seemed to be subscribing to a more traditional ideal of a self-sufficient economy: "encouraging our people to live at home; to stop sending outside the area money for food and feed and manufactured articles which can be more economically produced at home; to stop depending on a single crop which, for a number of years, has not returned the cost of production; to abandon the tenant system, which has proved unprofitable to both landlord and tenant."[100] And the Hugh Mac-

Rae Company's role as middle man in the purchase of hunting lands and the possible establishment of game reserves shows that they certainly did not envision a cultivated landscape without the possibility for that type of forest use. But its customers—buyers and sellers—were its fellow entrepreneurs and business associates, not common farmers.[101] Joseph Brown had shared with business associates such as Kinchen B. Council, James H. Chadbourn, and James Sprunt a strong attachment to the land they sought to transform.[102] In addition to owning and operating a farm near Whiteville,[103] he still enjoyed hunting quail with dog and gun, and wished at times to "try & completely forget that I have any business."[104] He could turn romantic and describe a "beautiful moonlight evening" as "just perfect for driving"; Conway, South Carolina had "a lovely river fine for boating for one who would like to work about four days each week & spend the other two on a small steam yacht cruising under the shade of grand old trees away from the whole world."[105] But this attachment was all the same informed by the attitude toward progress and improvement he shared with his circle of associates and friends. If Conway was a "quaint old town," the train trip there was nevertheless "a very disagreeable one . . . eat and sleep anywhere we can."[106] The project of improving the rugged backwoods, although underway, was clearly unfinished, and men such as Brown and MacRae intended fully to take charge of its completion.

The way reformers responded to what they perceived as dissent and recalcitrance makes clear that they had a comprehensive vision of rural improvement that left no room for traditional values and practices that might disrupt the advance of progress as they defined it. The experience of Jewish refugee Arthur Flatow in the Van Eden Colony is a case in point. When Flatow wanted to build a house, Alvin Johnson, president of the New School for Social Research and the Alvin Corporation that oversaw the settlement, told him that he could not build his own house. The corporation was to build houses for all settlers, and the company emphasized homogeneity over the specific needs of individual families: "they must all be made exactly on the same plan, finished exactly the same way." Supervisor H. M. Pinckney informed Flatow that he was also not free to set his own hours, telling him to "stick more closely to the work planned for you" and discuss any misunderstandings only "at the close of the work day or at noon." "Under no condition are you to leave your work to excite or disturb the other settlers during the work day," Pinckney wrote. Flatow continued to question the corporation's managerial practices, however, and asked Pinckney why his children were not allowed to work on the farm when "our children must work in order to help us make a new living . . . I beg you very much, that, if my daughter works for the corporation, that nobody is to tell her, how long she can work on a day . . . That is my child and I take care for her." In the end, Johnson had to intervene and scolded Flatow for the unapproved expenditures he claimed on the corporation, for being a bad businessman, and for thinking too much about himself: "You refuse to work except on your own terms when there is nothing else you want to do; and when you do work for the Corporation you waste your energies grumbling and arguing with other men, spoiling their usefulness too. I want you to stop thinking so much about yourself, and begin to think more about the common good of the community." When Flatow appeared finally to have come around to the corporation's vision of the common good, Johnson proved to be as paternalistic as the men who built the lumber industry around the Green Swamp in his attitude toward the common farmers and workers who needed to be brought in line with the business elite's plans for improvement: "To be quarrelling with you is very distressing to me. I regard all my settlers as my children, and it is sweet to live on good terms with one's children and sad to be on bad terms."[107]

* * *

The business community's paternalistic attempts to use the cutover as a springboard for a complete makeover of southern rural life, then, made little headway in the face of persistent traditional practices in a mixed economy of farm and forest, of market production and competency, as well as in the absence of the demographic revolution proponents of immigration hoped to instigate. If anything, the recent history of American and global agriculture has shown that the successful implementation of scientific, large-scale, industrial farming techniques have far more thoroughly rent the fabric of rural life than the practices it sought to replace. By hanging on to their woodland holdings, the farmers around the Green Swamp may therefore very well have helped steer cutover development away from primarily agricultural solutions to their perceived shortcomings. Instead, they steered the reclamation of the cutover toward reforestation when they intensified the cultivation of their improved lands and gradually withdrew their livestock from their woodlands: domestic animals have done terrible damage to forest environments; and if O. H. L. Wernicke's dream of a southern ranching empire had come true the growing cutover would most likely have remained a vast, treeless plain.[108] Farmers, by reducing their flocks and herds and keeping them out of their wooded acreage, contributed to the health of both the forest environment and the mixed rural economy. As a result, neither a republic of small farms nor an empire of ranches but new forests would come after the lumbermen.

Yet we must resist the temptation to proclaim victory for the moral economy. When both MacRae's intensive truck farming revolution and Wernicke's expansive sheep ranching coup failed to sweep the cutover, they and their associates tried equally hard to impose their vision of proper land development on the project of reforestation. Although Wernicke considered reforestation the domain of state and federal governments, not of the private companies that owned most of the cutover and could not afford to wait for the return of a profitable timber crop, he acknowledged the possibility and desirability of restoring the forest to its former glory and profitability. "Such reforestation," he said, "would . . . contribute greatly to the value and desirability of the remaining lands for grazing or other purposes."[109] MacRae's Associated Committees on Rural Development concurred, noting that farmers who deserted land "that ought to be in forest" did themselves and the land a favor because a "thoughtful and far-seeing policy of forestation for the South is highly desirable."[110] Around the Green Swamp, it turned out that more land than Hugh MacRae expected "ought to be" forestland and that the formulation of "thoughtful and far-seeing" policies would become yet another vehicle for imposing a capitalist vision of legibility on the landscape.

5

"It Would Seem Necessary to Restrict Such People"

"With the resettlement prospect out of the way," engineer Eric Norden wrote to Waccamaw Lumber Corporation president F. L. Finkenstaedt in May 1936, "it occurs to me that 'Now is the time for all good men to come to the aid of the National Forest.'" A month earlier, the National Forest Reservation Commission had approved, as part of the New Deal, the purchase of 18,430 acres for the establishment of five national forests in North Carolina. During the two preceding years, the commission had also expressed interest in a "Cape Fear Unit" as proposed by Norden and Finkenstaedt, who sought to include in a possible federal forest purchase the Board of Education swamplands and the Waccamaw Lumber Company's timberlands that they had failed to sell to settlement developers. According to Norden, Regional Forester Joseph Kircher considered the unit "the most attractive for forest reproduction of any in the records of the Forest Service." Indeed, the proposed reserve would go hand in hand with the nearby establishment of a federally supported pulp mill, which would draw on the reserve for its pulpwood supply. In addition, claimed Carl Taylor of the Agricultural Adjustment Agency, "the adoption of this Unit would go a long ways towards relieving distress and unemployment in Brunswick County and in the lower part of Columbus County," a dire necessity during the Great Depression. However, despite almost two years of lobbying by Norden, Finkenstaedt, Nelson MacRae, and others, the Lower Cape Fear area did not get to share the benefits of the $72,000 allotted for federal land purchases in North Carolina; as Kircher told MacRae in 1934, the state "already has more land included in National Forest Purchase Units than most southern states." Norden tried to keep the idea alive; but just as his appeals in person to the National Forest Reservation Commission prior to April 1936 failed to secure a national forest for the region, so did the engineer's subsequent call for private initiative.[1]

The attempt to convince the federal government to establish a national forest in North Carolina's southeastern corner was only one small step in a much larger, nationwide movement toward the adoption of conservation as a central tenet of land and resource use. Conservation—broadly conceived as the sustained use and regeneration of natural resources through the efficient, scientific management of their protection, cultivation, and use—had become an important issue on the federal level during the presidency of Theodore Roosevelt and remained central to economic policy during the first half of the twentieth century. In the face of much industrial waste and individual abuse of natural resources, conservationists sought to implement efficient methods of resource management, preferably government-supervised, in such crucial areas as public lands, waterways with power-producing potential, urban drinking water, and forests. Preaching what Samuel Hays called a "gospel of efficiency," they hoped to secure a well-managed, lasting supply of the natural resources that had fueled the prosperity of the Gilded Age and thereby avert the economic, political, social, and racial crises a shortage of exploitable and enjoyable resources might bring about.[2]

Earlier movements at the state level toward managed economic development through environmental protection and regulation had already established conservation as essentially an economic proposition that even in its design to establish game preserves and state parks mostly served capitalist interests.[3] North Carolina was somewhat of a vanguard state in preaching and practicing this type of environmental conservation, having employed such devoted conservationists as Gifford Pinchot, who would become the leading advocate of scientific forestry at the federal level, and W. W. Ashe, who shared with Pinchot the belief that conservation in general, and forestry in particular, served first and foremost an economic purpose.[4] Conservation, therefore, was not an environmentalist reproach of business, industry, or capitalism. As an ideology that developed toward the end of the nineteenth century in response to perceived threats to the *sustainability* of industrial capitalism, it served the goals of the dominant economic forces in forest industries and forest communities, especially when the severity and longevity of the Great Depression called into question the very supremacy of capitalism and galvanized governmental activism in resource management—a development that reflected New Dealers' belief that under-utilization of existing resources, not depletion, had greatly contributed to the economic slump.

As the work of Margaret Brown, Richard White, Joseph Taylor, and Karl Jacoby has shown for various historical and environmental contexts, the heavy-handed implementation of conservation programs by governments and their business allies often had detrimental effects on local communities. Whether it was the establishment of Great Smoky Mountains National Park and its celebration of invented Appalachian quaintness at the expense of vibrant mountain communities, the taming of Northwestern rivers at the expense of salmon migration and Indian fishing cultures, or the establishment of parks in the American Southwest at the expense of hunting, poaching, and squatting cultures, local traditions and their stores of informal knowledge fell victim to "progressive" government reform programs that sought to impose from above or outside a capitalist ideology detached from physical realities and ecological complexities.[5] Jacoby, especially, warns against interpreting and glorifying the rise of conservation as a promising step toward the curbing of big business' unbridled exploitation of the environment that Gilded Age governments had done so much to facilitate and encourage.[6]

But while conservation programs were certainly among the practical implementations of the increasingly dominant liberal notion that the triad of market, business-friendly government, and modern technology would eliminate biophysical limits on economic growth, neither the integrated national market nor the ideologies and institutions that underpinned it were simply the result of a revolution from above. Such programs also grew out of the deliberate expansion, on the *local* level, of markets, infrastructures, and resources, and their increased integration into a larger economy: the local always had a hand, often a large one, in its own transformation. This was especially true in states like North Carolina where progressive state governments initially had little land over which to exert any degree of real control. Unlike the federal government, which held sway over vast territories in the West, agents of the North Carolina Department of Conservation and Development needed the cooperation of the people who owned or controlled the available natural resources and the infrastructure that had developed around them. Only when a significant number of lumbermen, railroad companies, farmers, and woodland owners recognized that participation in the state's proposed conservation programs would serve their land use practices and increase the value of their stake in the economy were government conservationists able to implement their policies. Because this dependence fortified the position of local boosters, around the Green Swamp local proponents of progress came to embrace government programs to protect natural resources for economic, recreational, and aesthetic purposes.

After all, while it is important to remember that "local communities" are never homogeneous, the understanding homegrown entrepreneurs had of the people, economy, and environment in their communities was still "local" knowledge. The Green Swamp area was home not only to small and middling farmers but also to a fairly healthy lumber industry, an emerging pulp and paper industry, a dwindling but not yet insignificant turpentine industry, and a growing outdoors tourism industry, all of which continued to find extensive exploitable resources as a result of traditional farming practices and local patterns of woodland ownership. Proponents of sustained economic development became conservation advocates only when these various branches of the local forest economy threatened to encroach on one another and traditional petty use appeared more and more wasteful of nature's valuable resources during dire economic times. As the continued expansion of this mixed economy made increasingly volatile the relationship between the built environment, the natural environment, and their respective inhabitants, as one creature's forest activities were increasingly likely to infringe on another's, some system of management and arbitration to impose legibility and order became necessary. A government committed to economic and business development was therefore an attractive arbiter to those citizens with the biggest financial stake in an economically profitable relationship between humans and nature, while government programs drew on both private and public funds to sustain and renew the natural resources on which extractive and recreational industries depended for their survival.

At the same time, the move toward conservation obviously reflected neither a greater ecological appreciation for the ecosystems it affected nor a greater degree of communal solidarity; the conservation of forest and swamp through purchase and scientific management was a proposition shorn of sensitivity toward the complexity and diversity of natural and human community alike.[7] Instead, the endeavors of capitalist conservationists clashed with other local practices and beliefs, especially with those traditional methods of sustenance and land-clearing reformers fingered as the direst threats to nature and commerce: "excessive" hunting and fishing, "careless" use of fire, and any "wasteful" use of natural resources. The perceived need for resource conservation toward proper use in the face of intensified multiple uses and alleged abuses prompted a growing bureaucracy to pass a body of laws and policies toward that end. Thus drawing legal and ideological lines across forest and swamp, local proponents of conservation and their allies demonized traditional forest practices and restricted petty users' access to resources while reducing the diversity of ecosystems and communities to legible, manageable grids of usefulness. Rather than having "progress" imposed on their communities from above, they inscribed *their* understanding of the local community on broader visions of environmental and economic development.

"The former longleaf pine forest"

The Green Swamp region remained throughout the early twentieth century a green and wooded area that enabled forest industries to go about their business "with unabated energy."[8] The *Whiteville News Reporter* proclaimed in 1925, "The woods around Whiteville are full of all varieties of opportunities waiting for capital to come along and domesticate them"; as usual, entrepreneurs and boosters across the southeastern corner concurred.[9] There was still enough timber to support the area's extensive network of logging roads, small-to-medium sawmills, and other wood-processing establishments. Whenever trains passed through the region's small towns passengers could see one or more sawmills, lumberyards, or veneer plants close to the track. Even towns whose population did not exceed fifty souls could boast at least one locally owned sawmill.[10] Companies from outside the region also thought it worthwhile to set up operations in the woods of Brunswick and Columbus. In early 1924 the Beaufort Lumber Company purchased $1 million worth of timber in Brunswick County as well as an option on the Enterprise-Whiteville Lumber Company's mill in Whiteville, and it proceeded to construct a narrow gauge railroad from its mill in Fairmont to the Brunswick forests.[11] In 1925 the Jackson Brothers Lumber Company, one of North Carolina's largest lumber companies, moved into Whiteville, erected a mill town and a logging town nearby, and started producing lumber.[12] In 1928 the Tilghman Lumber Company of Dunn, North Carolina, purchased the Montgomery Lumber Company's timber holdings as well as its plant and logging equipment and erected a new mill in Fair Bluff.[13] The increased availability of the combustion engine also made it possible for smaller operators to move their mills further into the woods and facilitated a shift in transportation from trains to trucks. "The extensive and improved highway system," explained a state report, "has been responsible for profitable long distance truck hauling."[14]

Companies active in the growing southern pulp and paper industry also found the southeastern corner's woodlands enticing. After Charles Herty developed a chemical method to make paper from yellow pine pulp, previously considered unfit for pulp production because of its resinous nature, the southern pine belt started to attract a host of new investors and developers.[15] In the Cape Fear region, the Canal Wood Corporation of Conway and Wilmington and the West Virginia Pulp and Paper Company of New York were two major companies that purchased timber rights and tracts of land from lumber companies and individual landowners. The West Virginia Pulp and Paper Company operated in Columbus County primarily through its subsidiary, the Cooper River Timber Company, purchasing land from private landowners and from landowning companies. While West Virginia Pulp and Paper in 1924 declined the purchase of the Waccamaw Lumber Company's entire property when it saw the state of the land, in 1926 and 1927 the Cooper River Timber Company did buy from Waccamaw Lumber almost 4,000 acres close to the Atlantic Coast Line and sought to obtain a forty-acre tract fronting the railroad "in order to get an outlet for the timber already purchased."[16] The Canal Wood Corporation owned extensive lands and timber in Bladen, Columbus, and Robeson counties and annually shipped an estimated 40,000 cords of pulpwood to the International Paper Company's Georgetown mill.[17] The pulp and paper industry's standards for usable timber were less stringent than those of the lumber industry in terms of quality but more demanding in terms of quantity, yet they managed to sustain sizeable pulpwood harvests alongside the lumber industry's continued logging operations.

Even the turpentine industry, in spite of rapidly diminishing longleaf pine stands and stiff competition from producers further south, managed to survive in the southeastern corner. To be sure, North Carolina mostly proved an important repository of laborers with practical knowledge of turpentine extraction and processing as the industry shifted to South Carolina, Georgia, and Florida: "There is a common saying in Florida," wrote one industry advocate, "that all naval stores men come from Roberson [*sic*] County, North Carolina."[18] The industry's southward migration also reduced significantly North Carolina's share in production, a development that a diminishing supply of longleaf pine after years of boxing and logging could do little to reverse. The editor of the *Lumberton Argus* could only wax sentimental when he announced in 1903 "the passing of turpentine," depicting as "one of the most picturesque scenes of our memory . . . that of the negroes twisting 'rafting whips.'"[19] Still, scattered producers throughout the southeastern corner continued to tap their trees and run their stills; towns such as Shallotte, Seaside, Lockwood's Folly, Excelsior, Supply, and Leland continued to operate as small centers of production and distribution.[20] During the 1920s, several hopeful North Carolinians even prophesied the industry's imminent second coming in the Lower Cape Fear region, with the *News Reporter* going so far as to proclaim that "Turpentine May Again Bring Ships of the World to North Carolina Ports."[21] This revival never happened, but turpentine men did remain an economic presence in the forest—the last remaining still in the state of North Carolina closed its doors in Bladen County in 1942.[22]

A developing tourism industry and sportsman's culture used the forest economy's expanding infrastructure, especially the railroad network, for increased access to the outdoors, despite the decline in wildlife habitat and faunal diversity it had precipitated. Even if the mammals, birds, and fish that still inhabited southeastern North Carolina for the most part no longer roamed or swam in a pristine wilderness, they still sustained a local population of hunters and fishermen and made the region a popular resort for visiting sportsmen. Newspapers commented on the "hunting fever" and "good shooting" enjoyed by locals, who liked to visit newspaper offices to brag about their exploits and show off dead snakes and wildcat paws as trophies.[23] An amateur poet from Kentucky, whose connections with the Glanton Veneer Company in Whiteville had repeatedly brought him to the woods of Columbus, called the area "the hunter's paradise":

> For here you'll easily find
> Bear, deer, ducks, geese and turkeys
> And birds of every kind
> The rivers are full of fishes
> The lakes not far behind
> The swamps are full of bullfrogs
> The ocean full of brine.[24]

A "Plain Folk" song called "Fishing" asked, "What else can compare with just fishing?" "Not a thing," was the answer. Local fishermen rejoiced when two men "hooked" an eleven-pound specimen known as the "King of the Trout" from the Lumber River.[25]

The Green Swamp region's environment thus continued to sustain the identity of "forest communities," rooted not so much in the complexities of the longleaf ecosystem as in an understanding of nature's health in terms of the availability of harvestable trees and game. It continued to support the ways in which its human inhabitants had long appropriated forest and swamp for economic purposes and incorporated them into their lives. Yet many of the observations in the North Carolina Department of Conservation and Development's extensive 1928 timber survey indicated that the southeastern corner's forestlands, though still extensive, had nevertheless undergone severe, human-imposed changes. The department estimated that more than three-fourths of the area's total acreage consisted of "forest area," but little of it could be called pristine. Surveyor W. A. Peterson concluded that much of the merchantable timber standing in Robeson County was the product of "eighty percent of the forest . . . reproducing itself" after the timber harvest. He also observed that the annual cut on Columbus County's 489,662 forestland acres, an area he deemed "the heaviest timbered section of any of the counties" in his forestry district, was "being partially offset by the annual growth of timber on cut over lands." Brunswick County, too, was "essentially a timber producing county . . . as fully eighty percent of the forestland is covered with young growth in all stages from seedlings to timber." And Bladen County would top the list "in the number of acres devoted to the raising of timber . . . if proper attention was given" to protecting the county's 507,488 acres of forestland.[26] In addition, the four counties contained 89,487 acres of abandoned wasteland and 431,128 acres of forest or wild land that was not reproducing merchantable timber, much of which consisted of timberland that had sustained severe damage from logging and from subsequent fires. Finally, much of the 1.25 billion board feet of merchantable timber—which was confined to 19 percent of the forest area—consisted of loblolly pine that had replaced the original stand of longleaf pine. Of the latter there was only "a little": the economic relevance of the former "king among them all" now paled in comparison with that of gum, cypress, and oak.[27]

A statewide survey of the game harvest during the 1927–1928 season showed that while certain smaller species thrived in this patchwork forest environment, other species struggled to sustain their numbers. Almost 2,700 licensed hunters in the region reported killing 12,290 rabbits, 37,464 squirrels, and 44,984 quail, but only 201 deer and 65 wildcats.[28] Anglers enjoyed the fluvial plain's many creeks, lakes, and rivers, as well as the ocean on its edge; but they, too, found that increased human interaction with the environment had been detrimental to the health of certain fish species while others adapted quite well. Surveys showed that "undesirables," species such as catfish, blackfish, eel, and pike, were beginning to dominate the waters while "sport fish," which included shad, black bass, white perch, and goggle eye, declined in numbers. Between 1930 and 1931, for example, undesirables constituted 73 percent of all fish reported caught in Lake Waccamaw.[29]

"Vicious creatures that inhabit the jungles of this territory"

The distinction fishermen drew between sport fish and undesirables was not, however, a critical observation on the decline or disappearance of a unique ecological community, and it was more than just a sportsman's distinction between edible and inedible species, between hard-to-catch and easy-to-catch, or between classy swimmers and lowly bottom feeders. It reflected the increased categorization of nature in usable and non-usable components, as well as the classification of different resource utilization methods as "proper use" or "waste," both which distinctions gained in significance as forest communities altered and circumscribed a growing portion of their forestlands. After all, a natural area that could sustain such an array of extractive industries and activities was also still wild enough to provide shelter and breeding grounds for undesirable elements that threatened both the economic exploitation and the recreational enjoyment of nature. But when the geographic expansion of humans' extractive activities brought enterprising and sporting people in more frequent contact with wild or undesirable nature, the encounters increasingly underscored the latter's non-usability.

Although snake populations were most likely in fairly rapid decline as a result of human encroachment on the area's swamps and savannas, reports of "huge" rattlers and "monster snakes" were quite common and actually increased in frequency as time went on. These snakes met their inevitable deaths when they bothered hunters and anglers or ventured beyond the forests and swamps to visit farms and towns where their size, fangs, and rattles scared young and old alike.[30] Bears still roamed the region, especially the Green Swamp, and could sometimes be seen along the Waccamaw Lumber Corporation's logging railroad near Bolton. They, too, incurred the wrath of farmers and townspeople when they came out of the woods and ravaged crops, attacked people, killed livestock, or even "kidnapped" children.[31] An alligator emerging from swamp or stream would sometimes inspire a daring young man to ride the reptile's back for sport, but alligators generally frightened people into killing them, especially when they measured seven feet or more.[32] Sharks that ventured from oceanic waters into rivers and inlets, too, generally ended up as dead exhibits of human fear and ferociousness in the face of wild nature.[33] In the case of predatory birds such as eagles and owls, the threat they posed to domestic plumage and preferred game birds such as geese often signed their death warrants: a farmer in Mount Tabor received praise for killing a "monster hawk" that posed "a known terror to chickens and other native fowls."[34] The *Whiteville News Reporter* also reported the capture near Fair Bluff of a wildcat, "one of the . . . vicious creatures that inhabit the jungles of this surrounding territory."[35] In 1928 there was even a report of "a hyena or mystery varmint rambling Columbus county." This animal, whose existence some locals denied and others affirmed, had supposedly come up from Horry County, South Carolina, and killed fowl and housecats.[36]

The wilderness also sheltered unwanted human elements whose activities in the woods did not, in the eyes of many observers, constitute proper use of the forest's increasingly valuable resources and thereby obstructed proper development. The Green Swamp and its surrounding wetlands had always been popular hideouts for outlaws and other fugitives whose disappearance into the thickets had long fueled the popular swamp mystique. Well into the twentieth century, they remained places of refuge for those wanted by the law, as in the case of a Makatoka shooting suspect who took off for the Green Swamp.[37] Also disturbing was the behavior of some laborers at the various logging camps deep in the woods. Besides drinking illegal spirits, workers stole tools and feed from their employers and engaged in violence.[38] One lumber company employee participated in the abduction—into the forest—and shearing of a sixty-year-old man whose long hair had been an affront to a number of unknown men hiding behind the façade of the Ku Klux Klan.[39] Equally disconcerting was the gun violence among black lumber workers that sometimes threatened to turn "Black Bottom," the black section of the Longwood logging camp, into "No Man's Land." Such violence cost at least one black laborer at another Jackson Brothers lumber camp his life.[40]

Most notorious, however, were moonshiners who hid their stills and produce on the edge of swamps, in patches of dense woods and sometimes deep in the forest in order to avoid detection and capture, especially during Prohibition. The rural police force destroyed numerous illicit outfits and brought many a distiller to justice, much to the satisfaction of local moralists who objected most vehemently to illegal liquor making its way from producers in the forest to consumers in town through legal channels of commerce.[41] The *Lumberton Argus*, especially, railed against the almost openly defiant town of Hamlet, "the greatest hell-hole in all this section," as well as against the so-called corn licker express, the regular train to Robeson County that bootleggers were known to use for transportation of their wares under the not-so-watchful eye of the railroad's personnel.[42] Although newspaper editors no longer spoke of the "barbarizing" effect of too long a sojourn in the woods, they were very concerned with the rogue element that deliberately sought cover under the forest canopy and the increased frequency with which their activities spilled over into civilized society. Moreover, in an economy where the development of one legitimate extractive activity was increasingly likely to encroach on another—the crosstie business, for example, spelled doom for the turpentine industry, according to one observer—advocates of economic growth and proper resource use quickly grew impatient with ever-more flagrant recalcitrance and waste.

"Forest protection is like good roads"

As we have seen, these distinctions between "usable" and "non-usable" nature and between "proper" and "wasteful" resource use were not new to the early twentieth century. The forest industries that developed around the Green Swamp had always depended on a steady supply of wood from privately owned timberlands, and the land use ethic that valued equally the cultivation of farmland and the harvesting of its timber had long advocated exploitation but condemned "promiscuous" waste. It was not until the Progressive Era that this type of rhetoric gained widespread acceptance in state and national government circles and "proper use" advocates began to translate into actual conservation policy the idea that forests were a valuable resource in need of protection from wasteful exploitation. Anxiety about the economic consequences of an allegedly impending timber shortage spurred reformers into action, as did fear of the possible social consequences of the disappearance of the forest that had shaped the nation's frontier spirit. Under the national leadership of outdoors fanatic

Theodore Roosevelt and his head of forestry Gifford Pinchot, conservation became the watchword for natural resource management in urban-industrial America. As a result, forestry as a discipline gained greatly in stature among the various sciences to which Progressives turned for answers to the nation's many socioeconomic problems.

We should be careful, however, not to interpret the mainstreaming of conservationism as a major departure from earlier practices. Not only did conservation have its colonial and antebellum advocates, the tenets of progressive forest management were of a piece with the capitalist worldview that had driven at least partly successful attempts to make the landscape legible and to define "proper" and "improper" use within it. Extensively surveying timber stands to determine their potential yield in board feet and dollars; convincing private companies and farmers of the benefits of fire protection, prescribed burning, selective logging, and planting high-yield species; establishing national and state forests where governments could best implement and supervise such policies—all this added up to the implementation of *scientific* forest management that would sustain and expand both the older legibility of forests and, as Hays put it, their "greater commercial use."[43] Modern spoilers, therefore, like the wild tar-burners and renegades before them, were not a pest because they extracted forest resources but because they flagrantly disregarded the principles of what was now called "sustainable yield"—the ability of properly managed nature perpetually to renew its valuable gifts.

The experience of the Green Swamp region underpins Sherry Olson's assertion that the impending timber depletion, the evocation of which was a favorite scare tactic among reformers, was more myth than reality, that fears of a timber shortage were often unfounded.[44] Even during the height of the lumber boom, the area remained heavily timbered, and the natural offerings of both forest and cutover land continued to satisfy outdoor enthusiasts. At stake, therefore, was not the possible disappearance of timber and woodlands but the power to control and direct the extraction and proper use of resources in a maturing urban–industrial society where a growing number of well-established interests made demands on the existing resource base. The anxiety Progressives expressed, especially with regard to the extractive industries, was rooted in the so-called closing of the frontier, which gradually forced lumbermen and other agents of resource exploitation to abandon their migratory habits and pursue a strategy of resource renewal in a fixed location. This often meant that extractive industries, which had long been able to count on government support for easy, unsupervised access to untapped resources, had to accept the presence of a government that was rapidly taking on a more regulatory role. When settled communities pursued economic growth in a sedentary, extractive economy whose resources were abundant and renewable but not expandable, "waste" and "spoilers" became an ever-greater concern but now to the agents of a modern nation-state as well as to the business community and capitalist elites.[45] Operators who recognized the coexistence of different expanding industries as healthy diversification came to see industrywide cooperation between government and like-minded entrepreneurs as an appropriate way to protect resources from those who wasted them. Especially after the mobilization for World War I, the material demands which had forged closer, regulation-driven ties between business and government, the region's private and corporate advocates of "proper use" sought cooperative relationships with governments that might sustain greater commercial forest use.

The drive for such cooperation gained momentum when Progressive Era politics propelled into power many advocates of government activism who shared with local elites a desire for resource conservation through managed economic development, a preference for companies whose size or rootedness made them likely candidates for incorporation in regulatory programs, and a dislike of those individuals and small, migratory companies whose activities threatened natural and human-made environments alike. J. S. Holmes, for example, became North Carolina's State Forester in 1915 after having worked for the state's Geological and Economic Survey since 1909. Sharing his mother's "deep, intense love of nature as well as of God and her fellow creatures," and his father's devotion to natural science, he pursued an education and career in forestry. He devoted much of his professional life to forest protection through the establishment of a well-funded government structure that could forge cooperative ties with private forest interests.[46] As such, Holmes resembled the majority of post–World War I Progressives: their endorsement of business associations and government-supported private initiatives as the tools of progress, order, and efficiency after the uneven socioeconomic developments of the Gilded Age and the sudden surge in federal power during the Great War had by then fully subsumed religious beliefs regarding the care of nature and fellow creatures. "Forest Protection is like good roads, schools, the public health movement or the movement for better agricultural methods," one editor drove the point home.[47]

The North Carolina government thus became a rather partial arbiter in the growing conflict between those who propagated "proper" forest use and those forest dwellers whose different practices therefore constituted "waste." This deepening schism united under the government's conservation umbrella lumbermen who sought to sustain the harvest of trees, sportsmen who sought to sustain wildlife and its habitat for outdoors recreation, and anyone else who wanted to maintain both nature and an infrastructure for its continued exploitation. It pitted these interests against people whose supposedly destructive behavior wasted valuable timber and precious habitat, whose indiscriminate hunting and fishing practices threatened the perpetuation of harvestable species, and whose illicit, unproductive behavior, both inside and outside the woods, might give a town or region a bad reputation and scare off visiting sportsmen and potential investors.

Consequently, the North Carolina government began to question the methods of so-called peckerwood mills that, despite the lumber industry's "official" acceptance of selective logging methods through its representative organs, continued to clear-cut their holdings. Cutting small timber was an economic waste in two ways. First, the cost of cutting and hauling small trees tended to exceed the return on the amount of lumber they produced. Second, the cutting of young growth also endangered the future profitability of growing forests because small trees produced fewer seeds than those with a larger diameter. Larger companies, through their associations, had long agreed to refrain from cutting timber with a diameter of less than ten or twelve inches, but small operators tended not to subscribe to this policy. Echoing one of big lumber's main concerns, government agent W. A. Peterson charged that the "cutting of immature trees by small mill operators is common and should be discouraged." Counties like Robeson, he wrote, needed "conservation in the cutting of undersized timber . . . With proper protection to its growing forests and a little more thought given to present wasteful logging methods, [it] will become one of the largest timber producing Counties in the State."[48]

Progressives also managed better to protect forest resources from the ravenous hunger of free-ranging hogs when North Carolina lawmakers closed the open range in 1922, obligating farmers to fence in their hogs or face stiff penalties. Whereas proponents of agricultural reform considered the open-range razorback a staple of backward southern farming, conservationists directed their anger mostly at the hog's propensity to eat almost anything it rooted up and its particular fondness for longleaf pine seeds and seedlings. The stock law, which obligated farmers to pen in their livestock, was a victory for conservationists, who deemed the hog "an enemy of the pine."[49]

The North Carolina legislature's passage of beefed-up game and fishing laws strengthened the position of sportsmen and their allies who saw a serious threat to their interests, similar to that posed by peckerwood mills, in the behavior of people who killed small game and fish in such numbers that they jeopardized the survival of certain species. Many sportsmen were businessmen who harbored resentment toward people who hunted and fished for food or money because such activities contradicted the so-called sportsman ethos; according to assistant director J. K. Dixon of the Department of Conservation and Development, the ethic-bound sportsman "protects the pleasure of hunting."[50] The idea that staging a hunt and a pig barbecue was especially fine in timberland with "no underbrush . . . so you can see how to shoot" also indicated that a well-managed forest was preferable to a wilder one when it came to "a good time and good eats."[51] Hunters who killed for money or food tended not to consider the maintenance of species, thus compromising the sportsman's enjoyment of nature's bounty—when fur prices rose, for example, as happened in 1926, many hunters took as much game as they could.[52] Oscar High, a prominent merchant of Whiteville who facilitated the proper enjoyment of nature by renting out cottages at Lake Waccamaw to vacationing businessmen and Boy Scout troops, informed State Forester Holmes that in Columbus County "a number of men . . . were shooting nearly every day throughout the winter and it would seem necessary to restrict such people."[53] A similar complaint was heard from an inhabitant of Sessoms in Robeson County, who urged the editor of the *Lumberton Argus* to remind some of the townspeople—"the parties are well known"—to observe the deer hunting season and not mount a drive every few days.[54]

Concerned sportsmen applauded efforts on the part of government and law officials to keep the spoilers out, even if those efforts infringed on the livelihood and food supply of people who hunted for money and sustenance. The state, for example, used its power over lakes and other water systems as well as the surrounding land up to the high water mark to turn environments such as Lake Waccamaw, White Lake, Singletary Lake, and Jones and Salters lakes into game refuges where species could restore their numbers as the "killing or taking of game or other birds and animals upon or above any and all" of these lakes was prohibited.[55] The state also worked closely with the landowning elite of farmers, lumbermen, and other timberland owners who sold some of their cutover timberlands for game conservation purposes or counted on the state's commitment to protecting private property when they secured exclusive hunting privileges on private land as another solution to the problem of wasteful hunting.[56] Hunting and fishing seasons, in combination with limits on the number of game animals harvested and fish caught, further played into the hands of progressive sportsmen who championed the thrill of the chase rather than the largest possible harvest. The appearance of game wardens in the woods around the Green Swamp signaled the successful strides the alliance of capital and government was making toward progressive forest regulation. Tellingly, when the North Carolina government solicited feedback on the success of its hunting laws, it predominantly interviewed farmers and focused its survey almost exclusively on whether any hunters had been trespassing on private lands.

Prohibition was, in the same vein, the federal government's answer to the prayers of the Progressive Era's moral crusaders, who sought to banish from society the evils of alcohol consumption. Alcohol use was often the reason behind people's dismissal for "disorderly walk" from churches that policed the boundaries of orderly society but kept silent on the environmental ravages of business enterprise; many of the business elite were, after all, church leaders as well. J.P. Stephens' complaint that several of his fellow passengers on the train to the game lands of Robeson County "were loaded up with 'bug-juice'" was not an uncommon observation on the troublesome link between liquor and forest use.[57] Given the propensity of rural moonshiners for hiding their operations in the many swamps and forests that surrounded farms, the fight against alcohol production and consumption was, in effect, a significant part of forest regulation. The violence in the lumber camps was but one manifestation of disorder that could be traced back to liquor intake; reformers naturally objected to the production of alcohol in the camps' vicinity. But in the eyes of "proper use" advocates, moonshiners, who might claim theirs was a regular, if illegal, forest industry, also abused resources when they hid their outfits in the forests, utilized forest streams, or used the region's transportation network for their degenerate purposes. To Progressive reformers, moonshiners, like small mill operators, open-range hog farmers, and so-called game hogs, were enemies of an orderly, decent, progressive society—their uses of the forest threatened the resource base on which America's prosperity and character had been built.

"The forest fire menace"

As economic development and population growth strained the relationships between forest, farm, and town, it was the peckerwood mills, careless hunters, traditional farmers, moonshiners, and other alleged "spoilers" who were most closely associated with the element that, despite its long history as an indispensable tool of human progress, came to be seen as the forest's most destructive enemy: fire—also known as the "Red Demon" and the "Evil Spirit."[58] For communities that lived in close proximity to the forest and whose economies were inextricably tied to the woods, forest fires posed a very direct and intimate threat to both the commercial and the recreational side of forest exploitation. Fires killed wildlife, obliterated habitats, threatened the built environment on the edge of the forest, and left unsightly and unproductive stretches of charred land, all without giving anything of value in return. "The citizens of no community want its beautiful autumn landscape blotted out with smoke from burning woodlands," wrote the *Whiteville News Reporter*.[59] Neither did citizens appreciate the destruction of power lines, as happened in Bladen County when fire burned the Tide Water Power Company's power line poles and burning trees fell across the lines.[60] Most troubling to advocates of economic development, however, was the damage fire inflicted on stands of valuable timber. In fact, empowered by the methods of scientific forestry and driven by its ideology, they measured the impact of fire primarily in lost board feet and dollars, secondarily in physical damage to man-made structures, and only then in damage to the forest as an aesthetically pleasing, let alone ecologically significant natural environment. During the "most disastrous" spring of 1907, for example, fires across North Carolina burned over 750,000 acres of forestland and more than $4 million in property.[61] In the course of twenty-five years, fires destroyed an estimated $22 million worth of timber in Columbus County alone.[62] These same observers, consequently, came to look upon incendiary forest uses and their practitioners as antagonistic to the kind of proper resource exploitation required for economic growth.

Most leading causes of fire were directly related to the close proximity of economic and recreational activities to the forest environment. Activities and processes involved in three of the regional economy's mainstays—agriculture, lumbering, and railroad transportation—were directly responsible for quite a few fires in the area's forests. Farmers burning brush caused ten of the forty-two fires that swept Bladen County in 1916, while railroads caused two, and lumbering camps three.[63] Of the 115 fires that burned in Columbus County during a ten-month period between 1929 and 1930, twenty-seven were caused by brush burning, thirteen by lumbering, and two by railroads.[64] Brush burning was a common activity among many southern farmers, who annually set fire to their fields to increase their fertility or to stimulate the growth of new grass for their cattle. "Much new ground is being cleared by local farmers," the *News Reporter* depicted a familiar scene in January of 1928—"Fires burning at night show that the farmers are working over-time."[65] Lumber camp laborers, who depended for their livelihood on logging and milling operations but who had no direct stake in the capital timber represented, brought several fire hazards deep into the forest at once—they smoked, operated railroads and steam machinery, and cooked dinner on wood stoves. The engines that ran up and down the major tracks and the numerous logging railroads spat sparks into the dry growth along the track, despite being outfitted with protective metal plates that served to keep those sparks from flying.[66] J. A. McLean, a dealer in general merchandise in Fayetteville, had to report a "right bad forest fire" and a "rite [*sic*] bad . . . fite" when "Train #80 set the grass on fire" and burned 100 acres of sandhill woods in late 1922.[67]

Proponents of orderly economic modernization were once again quick to distinguish between fires that were unfortunate by-products of otherwise beneficial economic developments and fires that were the products of backward, callous, and illicit forest practices. Although they advocated caution and responsible behavior on the part of railroad operators and lumbermen, they considered railroads and lumber companies to be indispensable components of the developing forest economy and would never suggest that either cease their operations. They did, however, call on farmers to stop burning their woodlands and fields, which practice had for centuries served to clear land and temporarily improve soil quality. As with the issue of the open range, advocates of progressive change came to consider brush burning a backward tradition that did more damage than good. Farmers who burned "to ashes the grass and trees that Nature grew for us," the *News Reporter* asserted, neither increased the fertility of the soil nor killed the boll weevil, as popular belief would have it.[68] They did, however, pose a severe danger to the woods surrounding their farms as many a fire burned out of control and consumed numerous acres of forest.

In the same vein, progressive boosters depicted moonshiners, small sawmill operators, and unsportsmanlike hunters as major agents of fire destruction. At least one observer specifically identified "stilling whiskey" as a leading cause of forest fires. Portable steam sawmills posed a particular threat to the forest as their engines were operated much closer to the trees and underbrush than the central sawing units that larger companies usually operated in towns. Often these operators used fire to clear the forest floor and make room for the mill.[69] Small mills were also notorious for clear-cutting their limited timber holdings and leaving the cutover land, whose vegetation was much more prone to burn over than that of the original pine savannas, to spread fire to adjoining timber stands. "Bitter experience" also taught that hunters were "frequently careless with matches, smoking tobacco, and camp fires."[70] Like fishermen and campers, they were notorious for building campfires in the forest and leaving them to burn out by themselves. Often these fires consumed a significant stretch of forest with them as they burned out of control. One of the worst forest fires in the history of Columbus County, for example, "originated from the carelessness of hunters": In October of 1930, a forest area of 2,000 to 3,000 acres, situated between Chadbourn and Evergreen, burned for five days. According to the *News Reporter*, the fire confronted people "with a blinding and smothering smoke" that nearly obscured the sun; by the time the fire abated, "countless millions of feet of timber . . . had fallen under the forest fire menace."[71]

This explicitly hostile attitude toward specific causes and occurrences of fire constituted a significant shift away from the more ambiguous relationship that the inhabitants of the southeastern corner had forged with fire during the previous centuries; as such, it reflected the growing anxiety regarding the coexistence of multiple forest uses that accompanied the area's economic development. Fire was long more friend than foe to the people who settled the coastal plain, where white settlers quickly adopted the Native American practice of using fire for hunting, both to clear the forest floor of obstructive undergrowth and to drive game. The practice of clearing land and improving soil fertility through the burning of woods and brush survived well into the twentieth century, long after scientists had proven its failure to serve that purpose.[72] Fire also served to ward off mosquitoes in locations near swamps and other wetlands. Although fire tended to do substantial damage to human-built structures near the forest environment, often consuming entire sawmills, lumber yards, and turpentine distilleries, inhabitants of the Lower Cape Fear region commonly took the occurrence of forest fires for granted and waited for rain to put them out.[73] They generally expressed much more fear and concern regarding the violence of water and wind. Elizabeth Ellis Robeson of Bladen County, for example, casually observed a two-day "fire in the back woods" near her home but expressed consternation when a freshet in the Cape Fear washed out her fertile, cultivated bottomlands. "It is a great loss but we must be resigned to the dispensations of Providence," she wrote, ascribing to the destructive power of water a divine origin she did not necessarily see in fire.[74]

The widespread indifference toward forest fires in sparsely populated, heavily wooded areas explains the cavalier enforcement of a 1777 law and of subsequent legislation that prohibited people from setting fire to woodlands other than their own. The 1777 law required people to notify adjoining landowners two days in advance when they planned to set fire to their own property and obligated them to take "effectual care" to stop the spread of such fires to adjoining patented or vacant lands. But it proved hard to enforce since brush burning, the practitioners of which the law sought to curtail, was a common practice and neighbors were loath to bring charges against each other.[75] As a result of the enduring indifference toward forest fires, later legislation met with a similar fate. In 1905 the revised fire law sought to penalize "any wagoner or other person encamping in the open air" who left camp without fully extinguishing a campfire. This law sought to address the campfires of people in "the rougher and more remote parts of the State" where the absence of railroad transportation forced people to get their produce to market by way of long wagon trips that forced them to spend the night outdoors. In 1913 this section of the law was once again amended so as to include, explicitly, the campfires left by campers and hunters. Yet both versions of the law, aimed as they were at dealing with people in transit whose tracks often disappeared along with the forest they set on fire, proved as hard to enforce as the earlier laws.

However, continued population growth and economic development fragmented the forest to such an extent that fires, increasing in frequency as a result of the growing number of forest users, became likely to damage a growing number of adjacent timberlands, cultivated fields, or human-built structures. Although the exact turning point is unclear, this increase in the potential for fires to occur and inflict severe damage ultimately convinced local individuals interested in the continued exploitation of forest resources to pursue fire protection in conjunction with government agencies that had been advocating conservation since the 1890s. "Unless some provision is made to reduce the annual loss by fire and to increase the amount of timber grown in the forests of the state," wrote the boosters at the *News Reporter* in a telling word of warning, "North Carolina will . . . see the closing down of wood-working and related industries, representing many millions of dollars of investment and employing thousands of laborers."[76]

This embrace of a conservationist ethic on the part of economic interests was a welcome development for the leading advocate of fire protection, the Department of Conservation and Development's Forestry Division, which found that the success of protection programs hinged on the cooperation of forestland owners. Notwithstanding all the data on forest fires the department gathered from correspondents in the counties and published in government bulletins, the department did not own the resources it sought to protect; as long as the owners did not participate in some fashion, laws and information achieved little. In addition, the division did not own the monetary resources to fund an enforcement apparatus independent of the local economy. With local support and cooperation, however, the state was able to create a system of forest districts, overseen by district wardens. Each district consisted of several counties—Bladen, Brunswick, Columbus and Robeson, whenever they agreed to participate, were part of District No. 6—that were divided into areas overseen by fire wardens and deputy wardens. When county governments agreed to appropriate funds for forestry work and compensation for forest wardens and deputy wardens, the state and the federal governments matched those funds. The North Carolina Department of Conservation and Development then supplied the county wardens with firefighting equipment, aided in the distribution of informational materials, and helped implement protective measures, such as ditches and fire lines. Between 1 July 1929, and 1 May 1930, for example, Columbus County fire wardens posted 342 posters, interviewed

1,236 people, inspected 38 sawmills, visited 43 schools, appointed 122 deputies, and organized 20 motion picture exhibitions which drew a total of 5,080 people.[77] Besides these activities, the forest wardens patrolled their assigned areas and were in charge of rounding up people to help with the firefighting when they discovered a fire that appeared dangerous. Moreover, because the ignition of fires remained difficult to prevent but the early detection, containment, and extinction proved increasingly successful, the Department of Conservation and Development envisioned and initiated the erection of 125 fire towers in the district.[78]

The available evidence suggests how successful the cooperative structure was, both in terms of fire protection and the prosecution of fire offenders. During the six months of 1934 when Bladen and Brunswick did not cooperate, these counties suffered fifteen and nine fires respectively. These fires burned a total of 126,756 acres and caused $253,512 worth of damage. During the six months that they did cooperate, Bladen reported four fires and Brunswick fifteen, which represented not only a decrease in the number of fires but also a significant decline in the extent of fire-inflicted damage. The nineteen fires that occurred during the second half of 1934 burned a total of 5,465 areas and caused $12,910 worth of damage—a 94.6 percent decline in the number of burned acres per fire. Robeson County, which had protection the entire year, reported 237 fires but only 20,214 acres burned and $22,417 in damage. Columbus, unprotected the entire year, had 24 fires that burned 79,360 acres and caused damage worth $176,920. By contrast, during the ten months between 1929 and 1930 when Columbus did participate, 115 reported fires caused only $17,477 in damage.[79] The number of reported forest fires indicates that fires still started quite easily, but the drastic decrease in fire destruction shows that protection translated into quick containment of fires and a much smaller loss of acreage and stumpage. The new protective program also began successfully to prosecute offenders; people who wasted precious resources by setting fires suddenly found themselves faced with a much stronger warden apparatus. Telia McDuffey of Chadbourn, who sought to show her dislike of forest fire laws, burned 265 acres of private land when a warning about fire provoked in her a contrary reaction. Fire warden Minos Meares of Chadbourn Township brought her before the local magistrate, who fined her $20 and trial costs.[80] Seventeen year-old Sam McPherson found himself sentenced to three months of work on the Columbus County roads after he had set fire to a stretch of forest "just to see it burn."[81]

Such recalcitrance on the part of disenchanted individuals reflected the divisiveness of the fire protection issue within communities on the forest's edge, where fire remained an important tool of environmental exploitation. County governments often discontinued local funding of protection programs despite a growing awareness of their benefits on the state and national levels. Protection advocates, because of apathy on the part of county authorities, tried to bypass county officials through the organization of private companies into state or federally recognized cooperatives that would use private funds, matched by the state, to implement state guidelines for fire protection. Illustrative of this development was the joint effort of K. Clyde Council of the Council Tool Company, local fire warden Sidney Mears, and the local District Forester to convince the Columbus County commissioners to continue the fire protection program when the latter voted in May of 1927 not to renew the appropriation for fire prevention. While neighboring Bladen, Brunswick, and Robeson counties had all agreed to participate in the state's cooperative program, the commissioners' decision forced the Columbus County wardens and their deputies to return their firefighting equipment to the state.[82]

Throughout the early twentieth century, state conservationists mounted several efforts to avoid exactly such battles over what were usually meager county funds. On 11 November 1920, several lumbermen from the southeastern corner, including representatives of the Whiteville and Butters lumber companies, attended a conference in Goldsboro, where they heard State Forester Joseph Holmes and Chief Forest Fire Warden William Clark explain "the nature of . . . cooperative fire prevention work." Holmes and Clark unfolded a plan for fire protection "on which the lumbermen and the survey could unite." They proposed "to organize Forest Warden districts for patrol and educational work," with "the lumbermen and landowners cooperating with the state in the financial support of this work."[83] In the week following the conference Clark traveled through the southeastern corner to stir up interest in fire protection among lumbermen there; he met with representatives of the area's various lumber companies and land interests and "succeeded in interesting a number of [them] in the formation of a new district and the probable organization of an association to cooperate with the survey in fire protection."[84] On 25 August 1925, landowners and lumbermen—the Hugh MacRae Company, the Butters Lumber Company, the Jackson Brothers Lumber Company, and the Council Tool Company among them—met in Fayetteville to form the Timber

Owners Forest Protective Association, "an association for the protection of forestland from fire and for better private cooperation with the State Forest Service." Members agreed to pay one cent per acre of land, which the federal government would match and distribute among the counties according to the amount of land owned by the participants there. A total of 100,000 acres thus came under protection, a development fortified by the association's cooperation with the state in the construction of a ninety-three foot fire tower that enabled fire wardens to keep watch over a territory of 20,000 acres in the central portion of Brunswick County.[85] In 1928 the Cooper River Timber Company, Hugh McRae and Company, K. Clyde Council, Eric Norden, and the Waccamaw Lumber Company combined to form the Lake Waccamaw Forest Protective Association in Columbus County. This particular association proposed to build a ninety-three foot fire tower overlooking 200,000 acres of forestland and to engage in intensive forest fire control on a 30,000- to 50,000-acre area.[86] Once again strengthening the ties between business and government in the drive toward resource conservation, the Cooper River Timber Company sold to the Department of Conservation and Development, for the nominal fee of $1, the 15,000 square feet of land on which the tower was to be erected.[87]

Because local business communities played such a central role in the pursuit of fire protection programs, the protective structure as it came into being around the Green Swamp was really an extension of the business infrastructure, with the operation of the former serving the latter's greater objectives. Fire and forest wardens, watching the forest along roads and railroads, visiting sawmills and distilleries, and calling on locals to dig ditches and fight fires did not seek to shield the forest environment from human intrusion. Rather, they patrolled the increasingly blurred boundaries between the forest and the human-built environment in order to protect each from wasteful destruction by the other. Businessmen, committed to forest exploitation, headed the Timber Owners Forest Protective Association—at the organization's second meeting in January of 1926, its members elected R. I. Walter of the Butters Lumber Company president and F. B. Gault of the North Carolina Lumber Company as one of the directors.[88] The activities of the Council Tool Company perhaps best illustrate this interconnectedness between the various branches of the forest industry and conservation work. As a manufacturer of turpentine tools and an owner of timberland, Council Tool depended on a healthy forest to sustain its clientele and heartily supported forestry work. The company joined a number of fire cooperatives, and several of its directors served as fire wardens. The company also started to produce firefighting tools, such as fire rakes, fire swatters, and axes. It designed its fire rake—used to establish fire lines or breaks, roll burning logs, and dig, rake, or cut heavy grasses—with input from the North Carolina Forest Service.[89] In December of 1930 the company even donated a large trailer—complete with two fifty-five gallon water tanks, six fire pumps, eight rakes, six swatters, two bush axes, and twelve buckets—to the forest fire tower in Chadbourn.[90] For a small business whose directors and employees depended not only on profitable forest exploitation for their economic survival but also on fire protection for the sustenance of their lives in close proximity to the woods, forest management through private cooperation with state programs had become essential.

When Regional Forester Ralph C. Winkworth in the early 1950s conceived of "aerial delivery" in the Green Swamp to fight forest fires there, it was Jim Latty, vice president of the Riegel Paper Company in Bolton, who agreed to find the necessary land and construct an airstrip. Settling on an area called Bear Pen, which had solid enough earth to support airplane traffic, the company built a strip 6,000 feet long and 150 feet wide, with a million-gallon water tank at the 3,000 feet mark. To show its support, the North Carolina legislature agreed to purchase three Snow aircraft, each with a carrying capacity of 300 gallons, later replacing them with tankers with a capacity of 1,800 gallons. After the Forest Service persuaded other forest industry companies—such as International Paper, Georgia Pacific, Weyerhauser Paper Company, Canal Wood, and Acme Wood— to provide money for additional manpower and equipment, the Riegel Paper Company agreed to build a second runway. "It has served us well here in the Waccamaw Forest," the company observed, "which may very well be the largest high hazard, high valued pine forest plantation in all the United States."[91]

"Glad to save timber and property"

The close ties between the infrastructures of exploitation and conservation only strengthened when the Great Depression began to make itself felt in the southeastern corner, and the work associated with fire protection and resource conservation became an economic activity in and of itself. In the midst of an economic downturn that frequently forced lumber mills to shut down operations, the possible establishment of a national forest or a number of Civilian Conservation Corps camps was often the only hope of relief for young men living in communities on the forest's edge. Local leaders were quick to contact state and federal representatives and lobby for either form of government aid, eventually bringing several emergency and permanent camps to the region: a state forest camp for "junior white enrollees" in Elizabethtown that later converted to a private camp; a private woodland camp for "junior colored enrollees" in Bolton; a white camp "on the outskirts of Southport"; and a camp on the lands of K. Clyde Council in the Lake Waccamaw area.[92] Under the supervision of superintendents, enrollees erected telephone lines to improve emergency communication, constructed fire towers, fought forest fires, engaged in fire suppression work, conducted timber-type surveys, and built truck trails in order to make such remote places as the Green Swamp accessible to fire crews.[93] Aimed at the public good, work on private lands was not supposed to "increase the value of private property" and had to be "strictly in the interest of the public," but federal fire protection measures were of great advantage to local leaders with large stakes in protected timber resources.[94] With an economic upstroke not imminent, government-backed programs to protect the forest became the preferred way to employ young men, making the intertwined notions of economic development, resource conservation, government regulation, and "proper" employment and forest use the foundation for a policy of national recovery.

The *Backwoods Echo*, the newspaper of Company 4482 in the Elizabethtown camp, shows that fire suppression as propagated and practiced by government and business fit neatly into a larger capitalist worldview. Besides being put to work in field and forest, the boys in the CCC also received an education in the capitalist values that informed their conservation efforts. According to one of its contributors,

> The sight on which Camp White Lake is existing was . . . one of Gods [*sic*] most neglected creations . . . the sight on which it now prospers, was one of the most dreaded places in Bladen County. For years the hot sandy dust of this place had been absorbing the black soot of more than two destructive fires per annum . . . The Officers and boys . . . decided that this post of torment should be converted into a place where rational human beings would like to work and prosper rather than just carry on a drab existence.

The crew constructed "orderly, well laid out paths," "pleasant lounging places," a fish pond, and an "Oasis in the desert" consisting of "seats and ice water . . . confined under one artistic arbor." Emphasizing such qualities of proper cultivation as convenience, "equalized angels [*sic*]," and "perfect" construction, the boys in the CCC saw a "paradise . . . develop from waste before our very eyes."

Another contributor further emphasized that work, play, and learning in the camp greatly improved the boys' physical health and at the same time helped prevent them from turning to crime. The newspaper also contained pieces on Christian faith and character by Robert Livingston, as well as reports on the CCC boys helping to find a lost child from a nearby town, socializing new members into what initially struck one of them as a "convict camp," and being "glad to save timber and property" when fighting fires or building fire towers. Even more significant was their introduction to surveying and mapping, which taught them to think of land management in terms of monetary value, straight lines, and rectangular grids, with the nature they contained at best an afterthought—and even then identified according to "forest type" without regard for its ecological complexity:

> The boys in the section are being trained in the principles of the compass, chaining, plane table, and mapping . . . The purpose of the type mapping is to provide an accurate map to be used to gain knowledge of railroads, rivers, ponds, available water, fire breaks and fire trails in fire control forces. Also to show types of trees in different localities.

The map this project was to generate would "have the location, and type of country, of every road, creek and river when a fire breaks out"; as such, it would be "of great balue [*sic*] to thap[*sic*] part of the country."[95] This, of course, was cadastral legibility at its most advanced.

The practical experience and hands-on education that taught the young men of the Civilian Conservation Corps to appreciate the function and value of carefully managed lands in a larger socioeconomic context also found its way into a variety of sources from which American youth in general learned to value the combined efforts of government and progressive business. Popular author Zane Grey took a break from westerns with cowboys to write *The Young Forester*, a western with foresters that turned conservation efforts into acts of heroism in the Arizona forest, home to avaricious "lumber-sharks":

> No second glance was needed to tell that the lumbering here was careless and without thought for the future. It had been a clean cut . . . The stumps were all about three feet high, and that meant the waste of many thousands of feet of good lumber. . . . It was a wilderness of yellow brush, a dry jungle. . . . Fire must inevitably complete this work of ruin.[96]

Fire, of course, was "the forest's greatest foe," with coyotes, cougars, and bears adding to the danger. Fortunately, the young hero meets some old-time hunters whose store of practical knowledge and skills helps him navigate the treacherous wilderness and end up the possessor of a beautiful bear skin rug.

The Coca-Cola Company, eager to associate its beverages with enduring American values, issued a workbook-like pamphlet called *Lumber: Trees, Timber, Wood Products* for which youth could collect cards and glue them in designated spots. When put together, the words and pictures taught them the rich history and many present uses of lumber, the value of "Wood as Benefactor of Mankind," how "woodsmen protect American forests," and that "the first Colonists . . . drew from that forest the breath of Liberty." Young people, especially young campers, were exposed to Smokey Bear's exhortations to prevent forest fires, in part because "burned timber builds no homes." Boy Scouts and Girl Scouts, many of whom attended camps around Lake Waccamaw and the Green Swamp, belonged to organizations that saw knowledge and care of forests as essential to building character and instilling proper values. Taught the use of compass, measurement, and astronomy to master the navigation of wild places and at the same time realize the magnitude of "the enormous work of God," scouts would come to see "How the Trees Have Served Mankind," "What They Are Doing for Us Today," and "The Supreme Service Trees Give." They should strive to earn tree badges by helping plant trees and becoming knowledgeable about them, especially since the timber supply might soon depend on such forests and forest knowledge: "In your grandfather's day there was plenty of timber; in your grandchild's day there may be only man-planted forests." Therefore, the *Girl Scout Handbook* thundered, "*Be absolutely certain that you have extinguished your fire.*"[97]

In the worldview that underpinned the management of natural resources for sustained economic development, then, conservation, religion, public service, science, liberty, and prosperity went hand in hand. It is in this context, ultimately, that we must view the efforts of Federal Paper Board Company to publicize its reforestation efforts. Not only did the company provide oak trees to local schools, it sold surplus seedlings from its nursery "to landowners and other companies for reforestation of their own properties," supplied "many nearby residents . . . with a 'hand out' of about 25 seedlings for planting in their yards" during the seedling lifting season, and organized day programs for "4-H clubs, Future Farmers of America clubs, and other school oriented agriculture groups." Ultimately, this outreach served the greater, public good—as the company saw it—that subsumed the values and ideologies of "Scouts, church groups, civic clubs, schools, colleges, families, and other groups," all of whom stood to benefit from visits to the company nursery.[98]

"Some trouble"

The creation and management of a growing number of state parks further illustrates how conservation projects ostensibly aimed at the greater good and the public interest entailed the setting aside and careful cultivation of parts of the environment to facilitate specific uses, propagated by business and state interests and inspired by their particular visions of development and proper use—at the expense of other such uses, interests, and visions. Whereas game reserves and private hunting properties served specifically to protect game and habitat for the enjoyment of sportsmen, the construction of state parks entailed the setting aside of land and water specifically to facilitate the leisurely enjoyment of nature. As with game reserves, the Department of Conservation and Development used its power over water to create land areas it could protect and control, seeking to purchase lakeside lands from usually quite obstinate private owners. Salters and Jones lakes in Bladen County, for example, became the core of a state park when the department began purchasing the severely damaged and burned-over lands that surrounded them. "Not only has all reproduction been killed but a considerable portion of the standing trees have also been destroyed," wrote Holmes; yet with improved fishing and "a new growth of pine . . . a most attractive State Park could be established."[99] In the case of Jones Lake, the state received help from the U.S. government when the Department of Agriculture, using its power under the Weeks Law and

Figure 11. A view of Lake Waccamaw from Lake Waccamaw State Park. (Photograph by author, 2006.)

the Clarks-McNary Act, forced some unyielding landowners in Bladen County to give up their claims to some of these lands.[100]

State officials in such cases saw themselves as protectors of the common weal against landowners who sought to monopolize access to recreational areas: "It is the plain duty of the state to interfere where the rights of the public in the use of any public property may be endangered," wrote Holmes in 1929. The target of the government's wrath, W. A. Holder, claimed that he owned not only the land around Baker's Lake but also the lake itself; he "forcibly removed all boats from the lake," refusing "anyone permission to fish there unless they pay him for the right."[101] A year earlier the sudden attempt by G. F. Bullard to lay exclusive claim to lands surrounding Salters and Jones lakes, despite the fact that every summer the lakeside attracted "hundreds of boy Scout's camps" and "many people from different parts of the state," had similarly spurred the Department of Conservation and Development into action.[102]

Once the state secured ownership of the lands, the young men of the Civilian Conservation Corps aided in the planting of 90,000 cypress tree seedlings in the bays around Jones and Singletary lakes as well as 210,000 longleaf pine seedlings in the entire Bladen Lake forest area, further cementing the relationship between state and federal government in the landscape's transformation for conservation purposes.[103] The state also oversaw the construction of boating docks and other man-made structures that facilitated human convenience rather than ecological purity or diversity. And while the "immediate popularity" of Jones Lake's facilities for blacks—which enticed bathers to drive from Greensboro and Winston-Salem in "private cars, taxis, trucks, and trailers equipped with church pews"—may seem to demonstrate the validity of the department's state park vision for preserving public access to recreational facilities, that vision also prohibited a range of other commercial, recreational, and sustenance uses.[104] As public environments whose cultivated nature was aimed at offering the greatest good to the greatest number of people, parks could be as restrictive as they were democratic. Because their offerings were for recreational consumption only, logging, turpentine extraction, land clearing, and hunting for food or money were prohibited. Whatever logging did occur served primarily as a forest management measure and took place under state supervision. A fifteen-point list of regulations governing the administration of state lakes, in combination with a nine-point list of fishing rules, and a three-point list of boating regulations, further turned the enjoyment of watery environments into a rather heavily regulated affair.

Some of the regulations, such as those prohibiting littering and speed boating near bathing spots, made sense in terms of protecting public safety. Other rules, however, such as the prohibitions on night fishing and selling fish caught in the lakes, were direct blows to what used to be a way of life in the region.[105] What the department referred to as "some trouble with cattle ranging on the different tracts" was an obvious reference to the continued tradition of utilizing what used to be the open range; it came to a quick end "through the efforts of the area forester and ranger." Even blueberry picking required a permit, which, although "free of charge to local inhabitants," heightened the level of control the state exerted over forest areas. Perhaps most telling was the department's wish to "take care of" three families not in its employ who occupied houses on state property; by this the department meant the removal of these families and the destruction or remodeling of "those houses unfit for habitation." In an act reminiscent of the desire to remove black tenants' shacks from the landscape, the department had already torn down ten old buildings, "mostly shacks and unfit for use." In this case, however, it had obliterated the physical signs of traditional human habitation in the forest landscape in order to make room for more fleeting recreational uses of supposedly natural woodlands.[106]

Because the department depended so heavily on the cooperation of large landowners, its agents often had no choice but to take into account, if not defer to, the opinions of lumber company officials or influential citizens who had a vested interest in what they considered proper forest use. K. Clyde Council, for example, informed J. S. Holmes of his family's idea "that the Waccamaw Lumber Company's lands bordering on the lake shore and running back into the swamp would make an ideal preserve for deer." Holmes wrote to Cameron MacRae to ask his opinion on how best to approach the owner of the lands surrounding Black Lake about a possible sale. Whenever the state incorporated local knowledge into its conservation policies, it turned to the business community and social elite.[107] Similarly, the creation of a bathing beach for blacks that included two latrines and toilets, one pier, six fireplaces, sewage and water systems, and an 18,000-foot long power line, ostensibly "to relieve the congestion on certain days at Waccamaw Lake," in effect affirmed a white vision of proper social order and community relations.[108] The fact that black citizens had to wait until 6 July to celebrate Independence Day at the lake, while deeply ironic, was emblematic of the power structure that undergirded outdoors recreation.

Here was an extensive system of hunting and fishing licenses, restrictive sporting seasons, bagging limits, game reserves, state parks, fire districts, and fire towers, overseen by fish, game, and fire wardens from the upper crust of local communities, and representing an effort to manage and regulate flora, fauna, natural processes and the human footprint in and around the Green Swamp. This protective grid not only favored the goals and attitudes of conservationists and forest industry leaders over those of petty users whose traditional practices were now illegal, but it also helped bring about sufficiently extensive ecological changes so as to reshape the relationship between the communities around the Green Swamp and the forests that surrounded them. The effects of fire protection on the environment were manifold, but two were especially important to the way the people of the southeastern corner conceived of, and interacted with, the environment that surrounded them. First, fire suppression on both cutover and timberlands severely limited the regenerative capacities of "the pine that fire built."[109] The unique fire-resistant qualities of longleaf pine seedlings made the trees the central component of various types of longleaf pine communities whose many inhabitants had all adjusted to the fires that had swept the coastal plain for millions of years. In these communities, the longleaf pines depended on frequent fires for the suppression of other, less fire-resistant tree species that usually produced more seeds and that would quickly displace the longleaf pine as the dominant tree species if fire did not keep their growth in check. As a result of fire protection, human communities that had forged close ties with the longleaf pine forest found themselves having to adjust to a changing resource base when shortleaf pine, loblolly pine, various hardwood trees, and new undergrowth came to dominate second- and third-growth forests. Although these successive forests kept the region wooded and green, the unique longleaf pine communities largely disappeared, a loss felt not only by turpentine producers and longleaf pine lumber makers, but also by many now endangered indigenous plant and animal species, and by the people who enjoyed observing or hunting them.

Second, the suppression of natural fires that played such a central role in the maintenance of longleaf pine communities led to an accelerated buildup of undergrowth and grasses that, when allowed to proliferate, added greatly to the flammability of the forest ground cover. As a result, when fires did occur they were much more likely to burn out of control and do much more damage to the forest soil and to tree trunks, both of which tended to remain unharmed when flames leapt quickly across plant communities maintained by frequent fires. When fires burned out of control, partly as a result of conservation measures, they were much more likely to damage the precious natural resources and the valuable man-made structures that were at the core of the increasingly conservation-minded forest industry. To live with the increased flammability of their surroundings, people kept a much closer watch on the forest, fought harder battles with fires that did occur, and had to accept a management apparatus that looked with suspicion on their previous, less restricted relationship with the forest.[110]

In addition to fire suppression, conservation agencies introduced non-native animal species into the environment, measures that, although beneficial for the resuscitation of wildlife populations, reshaped not only the area's fauna but also humans' relationship with it by privileging the objectives of sport hunters over those of other forest users. The Department of Conservation and Development's introduction of non-native species such as Mexican quail was emblematic of a system of wildlife management that did not so much recognize the ecological role various animal species played in certain environments as it sought to sustain, even enhance, the place of wildlife in the scheme of human interaction with the environment as espoused by government agents and the business community.[111] For many forest users who were used to having ready access to the forest's faunal resources whenever they needed them, the environment that was being conserved in the name of sportsmanship was obviously not the same environment with which they had forged close economic and recreational relationships. Although State Forester Holmes warned of the "great danger of private interests securing all hunting rights" in Columbus County and testified to the "great advantage to the people" of public shooting grounds, the state remained dependent on the cooperation of those private interests to realize its conservation plans, which often meant the establishment of public preserves on lands that bore the unmistakable imprint of the forest economy.[112]

The water management that logically came with the establishment of game preserves and state parks on and around the lakes further strengthened the ties between a government devoted to planning and regulation and local elites in control of most natural resources—and revealed the interconnectedness of restricted use and ecological change. In order to circumvent Lake Waccamaw's recurring low water levels during the summer months and make "the lake much more popular for pleasure seekers, bathers, and fishermen," the state legislature granted local business leaders K. Clyde Council, Frances Beers Gault, and Oscar High permission to construct a dam and spillway at or near the lake's outlet to the Waccamaw River.[113] Although the grant specified that the dam should not obstruct the passage of boats or fish, its construction was of such a nature as to present an obstacle to "herds of eels" making their way from salty Caribbean waters to the fresh waters of Lake Waccamaw to spawn.[114] The dam, which successfully maintained a high water level in the lake, also caused problems for people living downstream who were confronted with lower water levels. A hydraulic engineer maintained that "cutting off the Lake Waccamaw flow by closing the dam is not working any appreciable harm to interest below the river" but suggested nonetheless that a water stage recorder be installed "in order to acquire data to contest any suit which might be brought against the state later."[115]

When the dam broke in July of 1926, assistant engineer Charles Ray informed J. S. Holmes that "the people in a certain section of Columbus County, known as Crusoe Island, were very much displeased" with the dam's construction, precisely because of the river's low water levels during the summer. Ray admitted to a lack of credible evidence but simply evoked the Crusoe Islanders' traditional reputation as renegades when he suggested that "the fact that they are very bitter in their feeling toward this dam" made it "possible that some one of them may have planted a charge of dynamite in the mud above the dam."[116] The sudden intervention in age-old water flows clearly caused friction and hostility towards the state's program and its private beneficiaries. Similarly, when a 1938 law prohibited "trawling for shrimp in any of the inland waters of North Carolina or in the Atlantic Ocean within three miles of the beach or ocean shore" between 1 April and 1 September, the mayor of Southport raised immediate objections. In a seaside economy that depended even more than inland communities on resource extraction from water as well as from woods, especially since the waters around Southport attracted commercial fisheries from outside the state, the law would "work a hardship on fifty or more men trying to make a living from this point." The Committee on Commercial Fisheries relented and allowed ocean trawling during the period covered by the law, but the remaining statute constituted yet another regulatory intervention in a traditional area of interaction between people and their environment.[117]

Finally, the movement toward the establishment of tree farms, while adding greatly to the region's stands of merchantable timber, including longleaf pine, constituted yet another blow to ecological diversity and the environment's facilitation of multiple forest uses. Encouraged by conservationists and the burgeoning pulp and paper industry alike, the production and conservation of merchantable timber and pulpwood was increasingly confined to tree farms and other carefully managed timberlands. Paper and lumber companies that chose not to abandon their logged-over lands began to adopt sustained yield policies and commenced planting endless rows of fast-growing trees, stimulating their growth through selective logging, fire protection, and the suppression of environmental diversity in favor of crop homogeneity.[118] The state and federal government, seeing in the establishment of these farms a kind of successful culmination of their campaigns to spread scientific forestry throughout the nation's woodlands, aided business interests in these efforts. North Carolina maintained a nursery that produced loblolly longleaf, shortleaf, and slash pine seedlings, as well as yellow poplar, black walnut, white cedar and cypress, from which it supplied both its state parks and private corporations.[119] A state inspection of the Mullins Lumber Company's lands near Fair Bluff in Robeson County revealed that the planting of 50,000 non-native slash pine seedlings had been a success—the company ordered another 40,000 seedlings "and would probably have taken more if they had been available at the nursery."[120] The state nursery also became a major supplier of seedlings to the Riegel Paper Company in Acme, a federally supported pulp and paper mill that had purchased the Waccamaw Lumber Corporation's 137,000 acres of cutover lands. Although initially dependent on local pulpwood suppliers, the company became largely self-supplying when it planted 100,000 pine seedlings on its vast holdings in 1939 and contracted for a million more for the following year. At the same time that state and federal employees thus moved trees into denuded environments, the Civilian Conservation Corps removed bushel upon bushel of pinecones out of its work environments, engaging in a curious exchange of carefully cultivated trees for seeds that could have facilitated natural reproduction and constituted a major source of food for local fauna.[121] Although certain types of game abounded on protected, reforesting cutover land to make managed timberlands attractive new hunting grounds for sportsmen, many other species associated with pristine longleaf pine communities did not flourish in the new, homogeneous environment.[122]

Because private business interests, public boosters, and state forestry officials pushed farmers also to embrace tree planting as a way to augment farming operations, the amount of forestland that could be classified as "pristine," "wild," or even "ecologically complex" continued to shrink. "Pines grow fast and make a clear profit on lands not suited for crops," wrote G. A. Cardwell, agricultural and industrial agent for the Atlantic Coast Line Railroad, who compared such timber to "bank collateral." The *Whiteville News Reporter*, in whose pages Caldwell's writing appeared, mounted nothing short of an editorial crusade to persuade Columbus County farmers of the value of timber as an agricultural crop, urging them to manage the stands they had and plant new ones on uncultivated tracts of land.[123] Similarly, the forestry division issued bulletins telling farmers how best to grow and market cordwood while managing "their woodlands on a permanent or long time crop producing basis."[124]

Because of the long history of timberland ownership among farmers, the movement to convince them to plant and manage trees as a profitable crop was more in tune with traditional environmental practices than the closing of the open range or the attempt to establish ten-acre truck farms. The *Progressive Farmer*'s suggestion to plant longleaf pines, if necessary "a dozen little seedlings in order to get one tree to live" so as to have "more of them around our homes," indicated once again that farmers valued the forests around them for economic, aesthetic, and domestic purposes.[125] Tree farming, however, elevated economic usage and legibility far above the other two, often sacrificing the aesthetic beauty and usage versatility of a diverse forest for the monetary benefits of a single crop. Austin Cary of Lake City, Florida, for example, praised in no uncertain terms the Southeastern Council's championing of forestry in the South, exalting

> an industry changed in its view of itself from early disappearance to anticipation of a future greater and more brilliant than its part ever was; men radically revising their operating methods of buying, protecting, and managing land with a view to future business . . . keen men are looking forward to a time . . . when vastly more timber will be available to work on than an industry of the present volume can utilize.[126]

But the creation and expansion of newly profitable woodlands from cutover or farmland spelled the destruction of habitat for plants, insects, lizards, birds, and other creatures for whom no sportsman voiced concern prior to the emergence of the modern environmentalist movement. As was the case with the establishment of fire-protected game preserves, refuges, and state parks, the cultivation of tree farms entailed the cordoning off of forestland for a mostly singular purpose, the often deliberate modification of the environment to enhance its utilitarian value for the forest economy, and the further erosion of the tradition of multiple forest uses that coexisted in the rich mosaic of wood and water that once dominated the coastal plain.

Epilogue

"Our Meat from God"

By the time the modern environmentalist movement emerged as the advocate for endangered flora and fauna, "wilderness," "biodiversity," "ecosystems," the "greening of America," and the idea that "small is beautiful" (as well as for human communities threatened by deforestation, soil erosion, air and water pollution, species extinction, and global warming), it faced more than recalcitrance and hostility from the corporate and national interests it vilified as the main culprits behind the exploitation, rape, and murder of nature. In areas like the Green Swamp region, environmentalists also faced the historical, traditional, and local environmental legacy of a community on the edge of swamps and forests and of communities that looked around and actually saw a lot of green, that considered themselves quite small already, and that knew wilderness intimately from experience. To these communities the sake of "ecosystems" and "biodiversity" seemed far less immediate than the necessities of providing for their daily bread and finding shelter from the nature around them, not to mention the desire to secure enough comfort and leisure to enjoy it. For a majority of people who knew the forests and swamps of the Green Swamp region intimately and experientially as inextricable parts of their communal history *and* progress, the ideology and practices of business and government for the most part affirmed their larger worldviews. Knowing forest and swamp culturally far more than they knew it ecologically, they continued to find in the orderly, corporately managed, state-governed, yet still abundant forest the economic, recreational, and spiritual sustenance they had long pursued. The joint efforts of businessmen and conservation-minded government agents to conserve forest resources around the Green Swamp for extractive and processing endeavors created a permanent grid of human intervention in the forest environment that imposed on the forest landscape a pattern of resource use that was as fragmented as the landscape itself appeared to be. But while the complexity and contingency of non-human nature could never be fully subsumed under this cadastral grid, and while creatures and elements and even human beings either were oblivious to or willfully ignored its straight lines, the grid did reflect the widespread conviction that industrious and progress-minded humans could and should bring both natural and human communities in line with their capitalist worldview.

In the forest and swamp communities of the Lower Cape Fear region, this conviction has translated into a landscape where modern environmentalists, now a powerful voice in American life, continue to find a lot of nature worth preserving and protecting for ecological reasons. The kind of communal life that developed there, a life of which ecologically diverse nature is an intimate and integral part, can therefore serve as an example for the mitigating effects of a community's shared culture and identity on business' frequent devastation of complex nature. Still, it is the very same drive to bring nature and people in line with the capitalist worldview, the very same dominance of business communities and their close ties with powerful nation-states, that has laid to terrible waste entire frail ecosystems, animal populations, and human communities, so we should not put too much stock in the alleged benevolence of the market and the nation-state, or in those who would trust them to provide for and defend the common good of the global—or the globe's—community. Nevertheless, modern environmentalists, even—or perhaps especially—where they have resorted to violent and destructive tactics, have not been able to displace entrenched notions that the business of America is business, or to halt the advance of urban, suburban, and commercial sprawl. "Humans reduce the natural world to 'landscape,'" wrote the late Warren Dean, lamenting the proliferation of "domesticated surroundings trimmed and shaped to fit some practical use or conventional aesthetic, or else, more chilling still, to 'space'—desert plains steamrollered and built upon to consecrate an extremity of species narcissism."[1] Bulldozers are even today rapidly leveling the forests around the Green Swamp to make way for "pastoral" golf communities and condominiums whose only references to the region's history as forest history are wooded perimeters that block our view of denuded grounds and idyllic names that evoke the forest that once was.[2] As the proliferation of technology, machinery, and human-built structures tells us, once in place, the social, economic, and political processes that depend on a widely shared understanding—among the people who, eagerly or reluctantly, participate in them—of material goods as good, and of more and cheaper goods as better, can and will inflict ever greater damage on life in the name of progress. When the market and the nation-state become the arbiters of good, truth, and right, as they have in the capitalist West, the aspects of nature and human communities that fall outside their narrow fields of vision suffer deeply.

As historian of religion Joseph Harontunian once dared write, "The good is not in 'goods.' The good is in justice, mercy, and peace. It is in consistency and integrity, in living according to truth and right. It inheres in men and not in things. It is other than the goodness of goods and without it goods are not good."[3] But men are not islands. Evolved or created, humans are *of* nature, *in* nature, and *with* nature, and so the good in men must inhere in that of which they are made as well—what Wendell Berry calls "the gift of good land." Aldo Leopold captured this notion in the land ethic he formulated when the economic prosperity created by World War II plunged the United States, driven by its "Abrahamic concept of land," into an orgy of consumerism and suburban sprawl: "We abuse land because we regard it as a commodity belonging to us. When we see land as a community to which we belong, we may begin to use it with love and respect. There is no other way for land to survive the impact of mechanized man."[4] Moreover, Leopold and Berry may have had a thing or two to say about Harontunian's implicit suggestion that goods *can* be good as long as men are good first. Harontunian may very well have thought that the goods we have *will* be good when they stem from our attempts to practice justice, mercy, and peace first. But too often the Ted Haggards and Joel Osteens of this world counsel their charges—from the church pew to the White House—with the precept that the rewards for inward grace come in the form of material goods that are always good, in spite of the willful disregard for the utter absence of justice, mercy, and peace in their production and consumption. As Aldo Leopold wrote about his attempts to rebuild "what we are losing elsewhere" by seeking a refuge from modernity on a Wisconsin sand farm: "It is here that we seek—and still find—our meat from God . . . Nothing could be more salutary at this stage than a little healthy contempt for a plethora of material blessings."[5] What the story of the Green Swamp and the "magic wand of capital" can teach us as we proceed to debate past, present, and future in ever more glaringly polarizing terms, is that neither what we have lost or are losing, nor what we have gained, is ever clear-cut. And if that seems like a bad pun with which to end a book about forest and swamp, it is nevertheless intended. We have something to learn from people who, like Leopold and the people living with the cutover, could create a nurturing, alternative way of living out of nature laid to waste by tapping into its remaining life—not an excuse to go on squandering nature, but a model of human resilience and adaptability, drawn from experiential knowledge and a mythological view of the world and our place in it.

Notes

Introduction: "The Magic Wand of Capital and Enterprise"

1. Advertisement for Federal Paper Board Company, Inc., from the *Whiteville News Reporter*, 1 October 1992, box 3, folder 270/3/1, "Industry," Lake Waccamaw History.

2. Richard Wiles, Jacqueline Savitz, and Brian A. Cohen, *Particulate Air Pollution: Human Mortality, Pollution Sources, and the Case for Tougher Clean Air Standards*, 3, 5, 20, 21. The only two polluters worse than Riegelwood were Duke Power's plant in Marshall and Carolina Power & Light in Roxboro.

3. Walter Gilman Curtis, *Reminiscences of Wilmington and Smithville-Southport, 1848–1900*, 55–57.

4. Janisse Ray, *Ecology of a Cracker Childhood*, 103, 123, 125.

5. Warren Dean, *With Broadax and Firebrand: The Destruction of the Brazilian Atlantic Forest*, 5.

6. Dean, "The Tasks of Latin American Environmental History," Harold K. Steen and Richard P. Tucker, eds., *Changing Tropical Forests: Historical Perspectives on Today's Challenges in Central and South America*, 5–15.

7. Gordon Wood, "The Enemy Is Us: Democratic Capitalism in the Early Republic," in *Wages of Independence: Capitalism in the Early American Republic*, ed. Paul A. Gilje, 137–153; quote on 137.

8. Dean, *With Broadax and Firebrand*, 116; also Richard Grove, *Green Imperialism: Colonial Expansion, Tropical Island Edens, and the Origins of Environmentalism, 1600–1860*, 3: "The available evidence shows that the seeds of modern conservationism developed as an integral part of the European encounter with the tropics and with local classifications and interpretations of the natural world and its symbolism." Shawn William Miller, *Fruitless Trees: Portuguese Conservation and Brazil's Colonial Timber*, blames the Portuguese Crown's restrictive forest policies for provoking settlers' spiteful destruction of trees.

9. Along these lines, Peter A. Coclanis, "Urbs in Horto," *Reviews in American History* 20, 14–20, deems William Cronon's *Nature's Metropolis: Chicago and the Great West* "a disturbingly anti-industrial, and ultimately, misanthropic book. For how else can we categorize a work that pooh-poohs the accomplishment of something so uncommon and unlikely in history as sustained economic growth?"

10. Throughout the book I refer to the area as the Green Swamp region, the Lower Cape Fear region, the southeastern corner of North Carolina, and the Inner Coastal

Plain Region. My main focus is on Brunswick, Columbus, Bladen, and Robeson counties with an occasional foray into Cumberland and New Hanover as two cities there, Fayetteville and Wilmington, played an important role in the development of the other counties.

11. U.S. Forest Service, *A Forest Atlas of the South* (1969), 14; Bladen and Brunswick were 75–100 percent forested, Columbus and Robeson 50–74 percent.

12. Stewart Ware, Cecil Frost, and Phillip Doerr, "Southern Mixed Hardwood Forest: The Former Longleaf Pine Forest," in *Biodiversity of the Southeastern United States: Lowland Terrestrial Communities*, eds. William H. Martin, Stephen G. Boyce, and Arthur Echternacht, 447–493.

13. The term "human footprint" has become increasingly common to capture the deep and lasting impact of human activity on the global environment. Eric W. Sanderson et al., "The Human Footprint and the Last of the Wild," *BioScience* 52, 891–904, write that "The human footprint is a global map of human influence on the land surface."

14. Advertisements for Council Tool from the *Whiteville News Reporter*, 15 October 1992, and the paper's centennial edition, October 1996, copies in box 3, folder 270/3/1, "Industry," Lake Waccamaw History.

15. Kirkpatrick Sale, *The Green Revolution: The American Environmental Movement, 1962–1992*, 6.

16. Curtis, *Reminiscences*, 61.

17. North Carolina *State Advertiser*, 8 July 1865: "When exhausted for Naval Stores," an advertiser wrote about a stretch of land west of Wilmington, "this property is very valuable for lumber . . . and when the timber is removed the land is believed to be very valuable for growing Cotton, Corn, Vegetables, Grapes, &c."

18. James Boyd, "Old Pines," ed. Richard Walser, *North Carolina in the Short Story*, 161–181.

19. I owe my interest in environmental history as well as a substantial intellectual debt to William Cronon, from whom I derive much of my terminology as well as my understanding of our common disciplinary objectives. See his definition of the three broad elements of environmental history in "Kennecott Journey: The Paths out of Town" in *Under an Open Sky: Rethinking America's Western Past*, eds. William Cronon, George Miles, Jay Gitlin, 28–51, especially 32. Equally important have been Cronon's insights into the importance of reckoning with cyclical, sacred time as much as with linear, secular time. See William Cronon, "The Uses of Environmental History," *Environmental History Review* (Fall 1993), 1–22.

20. *Whiteville News Reporter*, 3 April 1924 and 29 March 1928.

21. *Lumberton Argus*, 11 December 1902.

22. Wendell Berry, "The Use of Energy," in *The Unsettling of America: Culture & Agriculture*, 81–95. See also William Cronon, "The Trouble with Wilderness; or, Getting Back to the Wrong Nature," in *Uncommon Ground: Rethinking the Human Place in Nature*, ed. Cronon, 69–90, for a critical assessment of the "naturalness" of nature and the interconnectedness of the human and non-human aspects of landscapes; Richard White, "'Are

You an Environmentalist or Do You Work for a Living?': Work and Nature," in *Uncommon Ground*, 171–185, especially 173; and Simon Schama, *Landscape and Memory*, 9.

23. Susan Drake Emmerich, "Faith-Based Stewardship and Resolution of Environmental Conflict: An Ethnography of an Action Research Case of Tangier Island Watermen in the Chesapeake Bay," 2–3, 10, 11, 25, 41, 59, 204, 211.

24. Anne Raver, "Human Nature: A Georgia Daughter's Lullaby of the Pines," *New York Times*, 27 April 2000.

25. Ray, *Ecology of a Cracker Childhood*, 116–117.

26. Ibid., 89: Claude Levi-Strauss gave the name *bricoleur* to "folk recyclers, people of creativity, vision and skill who use castaways for purposes other than those originally intended, sometimes for art."

27. Ibid., 80.

28. The book that set the agenda for the study of capitalist transformation in a rural context is *The Countryside in the Age of Capitalist Transformation: Essays in the Social History of Rural America*, eds. Steven Hahn and Jonathan Prude. Most of the essays in that collection deal with transformations during the antebellum period and after 1865, not around the time of the American Revolution. Good introductions that cover the most important publications and that represent the various aspects of the debate on the colonial era and the early republic are Paul A. Gilje, "The Rise of Capitalism in the Early Republic" in *Wages of Independence*, ed. Gilje, 1–22, and Michael Merrill, "Putting 'Capitalism' in Its Place: A Review of Recent Literature," *William and Mary Quarterly* 52, 315–326. "Family security" is the term used by Richard Bushman, "Family Security in the Transition from Farm to City, 1750–1850," *Journal of Family History* 6, 238–256; "competency" was introduced by Daniel Vickers, "Competency and Competition: Economic Culture in Early America," *William and Mary Quarterly* 47, 3–29; see Bettye Hobbs Pruitt, "Self-Sufficiency and the Agricultural Economy," *William and Mary Quarterly* 41, 333–364, on self-sufficiency. In recent years historians such as Wilma A. Dunaway and Ronald L. Lewis have applied Immanuel Wallerstein's model of a capitalist world system to the history of the mountain South in order to place the development of rural communities in Appalachia, notorious for their alleged isolation, self-sufficiency, and backwardness, in the context of world economic developments. Contradicting the work of Jeanette Keith and other rural historians who identify self-sufficiency—"Making Do"—rather than market exchange as the defining feature of the rural economy, they argue that settlers to remote places deliberately maintained their ties with the market societies they left, rendering their new habitats capitalist from the get-go (Immanuel Wallerstein, *The Modern World-System* and *The Consolidation of the European World-Economy, 1600–1750*; Wilma A. Dunaway, *The First American Frontier: Transition to Capitalism in Southern Appalachia, 1700–1860*; Ronald L. Lewis, *Transforming the Appalachian Countryside: Railroads, Deforestation, and Social Change in West Virginia, 1880–1920*; and Jeanette Keith, *Country People in the New South: Tennessee's Upper Cumberland*).

29. Thompson introduced the concept in "The Moral Economy of the English Crowd

in the Eighteenth Century," *Past and Present* 50, 76–136; he assessed its impact in "The Moral Economy Reviewed," in E. P. Thompson, *Customs in Common*, 259–351.

30. James C. Scott, *The Moral Economy of the Peasant: Rebellion and Subsistence in Southeast Asia*.

31. See James D. Gwartney and Richard L. Stroup, *Economics: Private and Public Choice*, 7th ed., 6, 51.

32. Winifred Barr Rothenberg, *From Market Places to a Market Economy: The Transformation of Rural Massachusetts, 1750–1850*, 214.

33. Immanuel Wallerstein, *Historical Capitalism, with Capitalist Civilization*, 13–14.

34. See, for example, John Scott Strickland, who depicts the economy of newly freed slaves as a moral counterpoint to the rapid industrialization and harsh exploitation that characterized the southern economy after the Civil War in "Traditional Culture and Moral Economy: Social and Economic Change in the South Carolina Low Country, 1865–1910," *The Countryside in the Age of Capitalist Transformation*, eds. Hahn and Prude, 141–178. See also Joseph P. Reidy, *From Slavery to Agrarian Capitalism in the Cotton Plantation South: Central Georgia, 1800–1880*, 12–13: "Freedpeople resisted subjection to the kind of controls that landowners seemed anxious to enforce. They sought the relative autonomy of rental tenure or, failing that, opted for subsisting themselves through a combination of casual labor . . . and hunting and fishing. They were proletarians who resisted the full implications of proletarianization."

35. See Alasdair MacIntyre, *After Virtue*, 2nd ed., ix, x: "We have to learn from history and anthropology of the *variety* of moral practices, beliefs and conceptual schemes" (my italics). Indeed, "Marxism's moral defects and failures arise from the extent to which it, like liberal individualism, embodies the ethos of the distinctively modern and modernizing world, and that nothing less than a rejection of a large part of that ethos will provide us with a rationally and morally defensible standpoint from which to judge and act—and in terms of which to evaluate various rival and homogeneous moral schemes which compete for our allegiance."

36. Warren Dean, *With Broadax and Firebrand*, 5.

37. I thank David L. Carlton for helping me phrase and develop this insight into the market economy debate, discussed over coffee at Bongo Java in Nashville on 15 June 2006.

38. Elinor G.K. Melville, "Environmental and Social Change in the Valle del Mezquital, Mexico, 1521–1600," *Comparative Studies in Society and History* 32, 24–53.

39. Wood, "The Enemy is Us," 140.

40. See Brian J. Walsh and J. Richard Middleton, *The Transforming Vision: Shaping a Christian Worldview*, 17.

41. Milton Friedman, *Capitalism and Freedom*, 3, 4: "By relying primarily on voluntary cooperation and private enterprise, in both economic and other activities, we can insure that the private sector is a check on the powers of the governmental sector and an effective protection of freedom of speech, of religion, and of thought"; the book's "major theme is the role of competitive capitalism—the organization of the bulk of economic

activity through private enterprise operating in a free market—as a system of economic freedom and a necessary condition for political freedom." Friedman wrote in defense of what he deemed the proper meaning of "liberalism" following the appropriation of that term by those he considered its enemies, but given the current obfuscation of the term, especially in terms of the role of government in obtaining what are still the same fundamental objectives, I think that "capitalism," when understood as a comprehensive worldview, captures the common ground between liberals of every ilk much better. Among the most influential mega-church evangelicals are Joel Osteen and the recently disgraced Ted Haggard, the latter of whom writes: "I want my finances in order, my kids trained, and my wife to love life. I want good friends who are a delight and who provide protection for my family and me should life become difficult someday . . . I don't want surprises, scandals, or secrets . . . I want stability and, at the same time, steady, forward movement. I want the church to help me live life well, not exhaust me with endless 'worthwhile' projects." Haggard, *Dog Training, Fly Fishing, & Sharing Christ in the 21st Century*, 8. Jeff Sharlet, "Inside America's Most Powerful Megachurch," *Harper's Magazine*, May 2005, 41–54, writes: "One of Pastor Ted's favorite books is Thomas Friedman's *The Lexus and the Olive Tree*, which is now required reading for the hundreds of pastors under Ted's spiritual authority . . . From Friedman, Pastor Ted says he learned that everything, including spirituality, can be understood as a commodity. And unregulated trade, he concluded, was the key to achieving worldly freedom." According to Haggard, "pro-free markets" and "pro-private property" is "what evangelical stands for," so he seeks "to harness the forces of free-market capitalism in our ministry." Abraham Kuyper, *Lectures on Calvinism*, 40, wrote: "Only by Calvinism the psalm of liberty found its way from the troubled conscience to the lips . . . Calvinism has captured and guaranteed to us our constitutional civil rights . . . Simultaneously there went out from Western Europe that mighty movement which promoted the revival of science and art, opened new avenues to commerce and trade, beautified domestic and social life, exalted the middle classes to positions of honor, caused philanthropy to abound, and more than all this, elevated, purified, and ennobled moral life by puritanic seriousness." Kuyper's uncritical acceptance of commerce and trade as entirely compatible with Calvinist doctrine and not anathema to an elevated, purified, and ennobled moral life suggests that they worked together for a greater purpose. Ayn Rand's *Capitalism: The Unknown Ideal*, vii–viii, 19, is "a collection of essays on the moral aspects of capitalism . . . Our primary interest is not politics or economics as such, but 'man's nature and man's relationship to existence'—and that we advocate capitalism because it is the only system geared to the life of a rational being." Also, "capitalism is not merely the 'practical,' but the only *moral* system in history" (my italics). "Is man free?" asks Rand. "In mankind's history, capitalism is the only system that answers: Yes."

42. It is easy to forget in the digital age that behind the global exchange of capital and information in zeros and ones lies an environmental conquest that includes the generation and distribution of countless megawatts of electricity, the installation of miles of underground cable, and the launching and control of a satellite network in outer space—

not to mention the tremendous waste from discarded technology. See Richard White, "'Are You an Environmentalist or Do You Work for a Living?,'" 184–185; and Wendell Berry, "Why I Am Not Going to Buy a Computer" and "Feminism, the Body, and the Machine," in *What Are People For?: Essays by Wendell Berry*, 170–177, 178–196.

43. *Southport Leader*, 18 August 1892. Robert Stride, *Southport, North Carolina, and What the Completion of the Wilmington, Brunswick & Southport Railroad Means toward Its Future Development*, 2nd ed., 10–11.

44. Martin Bruegel, *Farm, Shop, Landing: The Rise of a Market Society in the Hudson Valley, 1780–1860*, 222, 226, 187–215.

45. Christine Meisner Rosen and Christopher C. Sellers, "The Nature of the Firm: Towards an Ecocultural History of Business," *Business History Review* 73, 577–600. A good example of the traditionally passive role of nature in business history is Alfred D. Chandler Jr., "Anthracite Coal and the Beginnings of the Industrial Revolution in the United States," in Thomas K. McCraw, *The Essential Alfred Chandler: Essays toward a Historical Theory of Big Business*, 307–342. Chandler attributed the slow development of American industry to the limited availability of coal east of the Alleghenies. Coal, in this essay, is simply waiting to be discovered and used; the processes by which coal is formed or the impact on the lives of miners and other laborers of the extraction and burning of coal are not a part of the story. Compare this to Elinor Melville, "Environmental and Social Change in the Valle del Mezquital, Mexico, 1521–1600" and *A Plague of Sheep: Environmental Consequences of the Conquest of Mexico*. Melville's work demonstrates "a close relationship between environmental change and economic growth and development" that "challenges the idea that the natural resource base acted simply as a passive constraint on development." See also David L. Carlton, "The Piedmont and Waccamaw Regions: An Economic Comparison," *South Carolina Historical Magazine* 88, 83–100.

46. The most immediately relevant collections of Berry's work are *The Unsettling of America*, *What Are People For?*, and *Sex, Economy, Freedom & Community*. The latter contains the essay "Christianity and the Survival of Creation," 93–116, in which he writes that "our destruction of nature is not just bad stewardship, or stupid economics, or a betrayal of family responsibility; it is the most horrid blasphemy."

47. Douglas Dowd, *Capitalism and Its Economics: A Critical History*, 2, bemoans "the several billions who have been . . . uprooted from their traditional ways of life as a result of capitalism's conquests."

48. Benedict Anderson, *Imagined Communities: Reflections on the Origin and Spread of Nationalism*, rev. ed., 6.

49. Berry, "Christianity and the Survival of Creation," in *Sex, Economy, Freedom & Community*, 93–116.

50. See, for example, Alberto Manguel, *Into the Looking-Glass Wood: Essays on Books, Reading, and the World*, 116: "No single individual is at the root of evil, but rather the collective, all-encompassing social structure that determines who is in and who is out, defining itself by that which it excludes."

51. See Margaret Stacey, "The Myth of Community Studies," *The British Journal of So-*

ciology 20, 134–147; Brett Williams, "The Concept of Community," *Reviews in Anthropology* 31, 339–350; George Giacinto Giarchi, "Caught in the Nets: A Critical Examination of the Use of the Concept of 'Networks' in Community Development Studies," *Community Development Journal* 36, 63–71. According to V. Thomas Parker, "The Community of an Individual: Implications for the Community Concept," *OIKOS* 104, 24–34, the use of concept of "community" in ecology is often equally uncritical—in many ways ecologists, too, take as self-evident the boundaries they seek to define and police.

52. Jonathan Kusel, "Introduction," in *Forest Communities, Community Forests,* eds. Jonathan Kusel and Elisa Adler, xv–xxi.

53. William R. Burch, Jr., "Foreword," Kusel and Adler, *Forest Communities*, vii–xii.

54. Sam Burns, "Catron County, New Mexico: Mirroring the West, Healing the Land, Rebuilding Community," Kusel and Adler, *Forest Communities*, 89–115.

55. Barb Cestero and Jill M. Belsky, "Collaboration for Community and Forest Well-Being in the Upper Swan Valley, Montana," in Kusel and Adler, *Forest Communities*, 149–169.

56. Peter Lavigne, "Revolutionizing County Forest Management in Minnesota: Aitkin County and Smartwood Certification" in Kusel and Adler, *Forest Communities*, 55–85.

57. Kusel, "Introduction," xv–xxi; my italics.

58. I am very much inspired by the writings of William T. Cavanaugh, who writes the following in "Killing for the Telephone Company: Why the Nation-State is not the Keeper of the Common Good," *Modern Theology* 20, 243–274: "In the public forum both sides of the NAFTA debate asked 'Will this treaty be good or bad for America?' Only a few marginal voices on the losing side were able to suggest that NAFTA would be good for some Americans and bad for others, namely, good for capital and bad for labor. Claims for the interests of groups must be justified in terms of national interests, but the wealthier classes are far more effective at presenting their interests as being national interests."

59. James C. Scott, *Seeing Like a State: How Certain Schemes to Improve the Human Condition Have Failed*, 2, 6, 8, 11–52.

Chapter 1. "A Country But Wild and Imperfect"

1. Lawrence S. Earley, *Looking for Longleaf: The Fall and Rise of an American Forest*, 38–39.

2. See, for example, *The Rand McNally Road Atlas* (2001), 75.

3. Laurence C. Walker and Brian P. Oswald, *The Southern Forest: Geography, Ecology, and Silviculture*, 2–6, 11–14, 55–58; G. Frederick Schwarz, *The Longleaf Pine in Virgin Forest: A Silvical Study*; U.S. Department of Agriculture, *Silvics of North America, Volume 1: Conifers*, 405–412.

4. Quoted in Michael P. Schafale, *Inventory of Longleaf Pine Natural Communities in North Carolina*, 6.

5. A blackwater river is a river with a deep, slow-moving channel that flows through

forested swamps and wetlands. Its color, ranging from clear tea to coffee, results from the leaching of tannins from the decaying leaves of adjoining vegetation. Brownwater rivers tend to be longer and therefore carry high loads of clay and silt, giving the water a brown coloration.

6. Richard LeBlond, *Inventory of the Natural Areas and Rare Species of Brunswick County, North Carolina*, 9. These brownwater landform features include channel bars, point bars and ridges, levees, oxbows, active and relict sloughs, and active and relict swales.

7. The phrase "a sloshy union" is from *The State* (Columbia, S.C.), 7 May 2000. For an insightful discussion of pocosins—and a lamentation for the growing scarcity of q's in our world—see Jack Temple Kirby, *Poquosin: A Study of Rural Landscape and Society*, especially xii.: "Swamps-on-hills are sometimes covered with water, but the water stand still, gradually seeping downward; and of course they are higher than swamps. Pocosins . . . are the middle landscape between swamps and the higher ridges on which humans build farms and towns."

8. Michael G. Messina and William H. Conner, eds., *Southern Forested Wetlands: Ecology and Management*.

9. Dirk Frankenberg, *The Nature of North Carolina's Southern Coast: Barrier Islands, Coastal Waters, and Wetlands*, 211.

10. See Richard Schweid, *Consider the Eel*, and Coastal U.S. Fish and Wildlife Service, *American Eel—Species Profiles: Life Histories and Environmental Requirements of Coastal Fishes and Invertebrates (North Atlantic)*. The American Eel spawns in the Sargasso Sea south of Bermuda (the oceanic phase) and during its life cycle makes its way to the eastern coastal waters of North, Central, and South America and inland up the estuaries and rivers (the estuarine and riverine phases). It lives in warm brackish and freshwater streams, estuaries, and coastal rivers, and sometimes in cold freshwater trout streams in mountainous regions before making its way back to the Sargasso Sea to spawn. Its abundance in North Carolina's many inlets, estuaries, and coastal rivers prompted John Lawson to exclaim that "Eels are no where in the World better, or more plentiful." See also Joseph V. Siry, *Marshes of the Ocean Shore: Development of an Ecological Ethic*, 3–4; Frankenberg, *The Nature of North Carolina's Southern Coast*, 204–213. Timothy Silver, *A New Face on the Countryside: Indians, Colonists, and Slaves in South Atlantic Forests, 1500–1800*, 15.

11. Ray, *Ecology of a Cracker Childhood*, 37–38, 151–156, 217–222.

12. See Richard Eston Schneider, "The Effect of Variation in Season of Burning on a Pine-Wiregrass Savanna in the Green Swamp, North Carolina," i–ii, 2–6. The occurrence of fire in different seasons has a diversifying effect: fires in the spring increased production more than fall or summer burns, although grasses had greater production following spring fires, shrubs after fall fires.

13. LeBlond, *Inventory of the Natural Areas and Rare Species of Brunswick County*, 12–14, 43.

14. Frankenberg, *The Nature of North Carolina's Southern Coast*, 200.

15. See Russell L. Kologiski, "The Phytosociology of the Green Swamp, North Carolina," *North Carolina Agricultural Experiment Station Technical Bulletin*, no. 250; Abigail Rome, "Vegetation Variation in a Pine-Wiregrass Savanna in the Green Swamp (North Carolina)."

16. Division of Environmental Management, *A Field Guide to North Carolina Wetlands: Report No. 96–01.*

17. Alan S. Weakley and S. Kay Scott, *Natural Features Summary and Preserve Design for Carolina Bays in Bladen and Cumberland Counties, North Carolina*, 4–5.

18. LeBlond, *Inventory of the Natural Areas and Rare Species of Columbus County, North Carolina*; Rebecca S. Marty, "The Floodplain Plant Communities of a Blackwater River: The Waccamaw River, North Carolina." S. Leonard, *Inventory of Populations of Thalictrum cooleyi and Its Occurrence Sites in North Carolina: Report to the North Carolina Natural Heritage Program.* Among the rare fish and mussel species in Lake Waccamaw are the Waccamaw darter, the Waccamaw silverside, the Waccamaw killifish, the broadtail madtom catfish, the Carolina pigmy sunfish, the Waccamaw fatmucket mussel, the Waccamaw lance pearlymussel, and the Waccamaw spike.

19. "What's So Special about the Green Swamp in N.C.?" http://www.geocities.com/greenswamp211/links.htm (accessed 26 July 2006); my italics.

20. See Vanessa Grigoriadis, "The Rise and Fall of the Eco-Radical Underground" *Rolling Stone* (July 2006). Demonizing radical environmentalists as "terrorists" has given the American government yet another reason—and additional funding—to wage war on their anti-capitalist activities. See also Will Baker, *Tony and the Cows: A True Story from the Range Wars*, 61–69.

21. Carolyn Merchant, *The Death of Nature: Women, Ecology and the Scientific Revolution*, xvi.

22. On the persistence of supernatural beliefs in the face of both strict religiosity and scientific developments see David D. Hall, *Worlds of Wonder, Days of Judgment: Popular Religious Belief in New England*, 16–20.

23. Ann Vileisis, *Discovering the Unknown Landscape: A History of America's Wetlands*, 33–37; Roderick Nash, *Wilderness and the American Mind*, 3rd ed., 13–22.

24. I derive my understanding of historical attitudes toward nature as *mythological* from Bronislaw Malinowski, who defined myth as a story "told in satisfaction of deep religious wants, moral cravings, social . . . requirements. Myth fulfills . . . an indispensable function: it expresses, enhances, and codifies belief; it safeguards and enforces morality; it . . . contains practical rules for the guidance of man. Myth is thus a vital ingredient of human civilization; it is not an idle tale, but a hard-worked active force; it is not an intellectual explanation or an artistic imagery, but a pragmatic character of . . . faith and moral wisdom, . . . a statement of a primeval, greater, and more relevant reality, by which the present life, fates and activities of mankind are determined." See Bronislaw Malinowski, *Myth in Primitive Psychology*, 19, 30. See also 47–50 Robert Inchausti, *Subversive Orthodoxy: Outlaws, Subversives, and Other Christians in Disguise*, for the persistence (and necessity) of a "countermythological" outlook on life in the modern world. Erich Auerbach,

Mimesis, 156, uses the phrase "the figural interpretation of history" to get at the intersection of myth and history: "every occurrence, in all its everyday reality, is simultaneously a part in a world-historical context through which each part is related to every other, and thus is likewise to be regarded as being of all times or above all time."

25. "The Legend of the Lake," typed manuscript, box 270/1/11, "Fiction/Poetry about Lake Waccamaw," Lake Waccamaw History.

26. J. Ronald Eyton and Judith I. Parkhurst, "A Re-evaluation of the Extraterrestrial Origins of the Carolina Bays," Geography Graduate Student Association, University of Illinois at Urbana Champaign, Paper No. 9 (April 1975); a copy can be found Lake Waccamaw History collection of the High Family Papers.

27. Connie Nelson and Floyd Harris, *The Film Junkie's Guide to North Carolina*, 113–114; my italics.

28. http://ils.unc.edu/parkproject/visit/jone/home.html (accessed 12 August 2007); my italics.

29. "The Riddle of Shallotte Inlet," in Charles Harry Whedbee, *Legends of the Outer Banks and Tar Heel Tidewater*, 3–6.

30. Bland Simpson, *Into the Sound Country: A Carolinian's Coastal Plain*, 161–162.

31. Minnie McIver Brown, "Uncle John Sutton: A Romance of Slavery Days," unpublished manuscript, folder 69, Brown Family Papers.

32. Ann Courtney Ward Little, *Columbus County, North Carolina: Recollections and Records*, 299–300.

33. This practice of partly deriving one's identity from one's proximity to danger and depravity is similar to slumming, being down with gangsta rap, proudly claiming one's lower class roots after rising above them, and romanticizing southern poor white culture. See, for example, Nathan Irvin Huggins' discussion of Carl Van Vechten's infatuation with the exotic danger of Harlem in *Harlem Renaissance*, 93–118, or Bruce J. Schulman, *The Seventies: The Great Shift in American Culture, Society, and Politics*, 102–117, on "redneck chic."

34. A Traveller, "Turpentine," *Southern Cultivator (1846)*.

35. *"A Visit to the Cape Fear in 1734," in James Sprunt's` Chronicles of the Cape Fear River, 1660–1916*, 2nd ed., 40–45.

36. Frances Anne Kemble, *Records of Later Life*, vol. 1, 183.

37. Una Pope-Hennessy, ed., *The Aristocratic Journey: Being the Outspoken Letters of Mrs. Basil Hall, Written During a Fourteen Months' Sojourn in America, 1827–1828*, 204; Herbert B. Adams, *The Life and Writings of Jared Sparks: Comprising Selections from His Journals and Correspondence*, 442; "Life and Travel in the Southern States," *Great Republic Monthly* 1, 80–84.

38. Catherine Eliza Clitherall Diaries, Series 1, vol. 4, 57.

39. Ibid., Series 1, vol. 5, 14, 16.

40. Reprinted in the *Carolina Observer*, 14 May 1829.

41. On the sublime, see Cronon, "The Trouble with Wilderness," especially 73–76; Nash, *Wilderness and the American Mind*, 45–46 and 54–65; and Andrew Wilton and

Tim Barringer, *American Sublime: Landscape Painting in the United States, 1820–1880*, 11–37.

42. See Genesis 1:28 (King James Version).

43. Barbara Novak, *Nature and Culture: American Landscape and Painting, 1825–1875*, 3.

44. Carol Sheriff, *The Artificial River: The Erie Canal and the Paradox of Progress, 1817–1862*, 32–35, writes that "By supplying its own waterways and a break in the mountains, nature offered humanity a tantalizing challenge to finish nature's work."

45. On the pastoral, see Leo Marx, *The Machine in the Garden: Technology and the Pastoral Ideal*, especially 3–11; Nash, *Wilderness and the American Mind* 6, 9–10, 30–33; and Novak, *Nature and Culture*, 170–173.

46. Clarence J. Glacken, *Traces on the Rhodian Shore: Nature and Culture in Western Thought from Ancient Times to the End of the Eighteenth Century*, 427.

47. Walter Gwynn, undated letter, Eric Norden Collection.

48. Clitherall Diaries, Series 1, vol. 4, 58.

49. Pope-Hennessy, ed. *The Aristocratic Journey*, 204; Edmund Kirke (James R. Gilmore, pseudo.), *Among the Pines: Or, South in Secession-Time*, 227–252, especially 230–231 and 241; Frederick Law Olmsted, *A Journey in the Seaboard Slave States, in the Years 1853–1854, with Remarks on Their Economy*, 388; "The Piney Woods," *De Bow's Review* 12, 361–369.

50. Adam Hodgson, *Letters from North America, Written during a Tour in the United States and Canada*, vol. 1, 37–38. This particular letter is dated 19 February 1820.

51. Nathaniel H. Bishop, *Voyage of the Paper Canoe: A Geographical Journey of 2500 Miles, from Quebec to the Gulf of Mexico during the Years 1874–5*, 218–227.

52. John Martin Hammond, *Winter Journeys in the South*, 201–207.

53. Williams Haynes, "Through a Jungle to the Old South," *Outing* 65, 478–486, and "Down the Old Lumbee," *Field and Stream* 23, 5–7 and 48–49, the latter being a later version of the same story; my italics.

54. Bland Simpson, *Into the Sound Country*, 170; my italics.

55. William Gordon to the Secretary, 13 May 1709, in William Laurence Sanders, ed., *The Colonial Records of North Carolina* (CRNC), vol. 1, 715.

56. "Report of Commissioners Sent from Barbadoes to Explore the River Cape Fear, in 1663," *CRNC*, vol. 1, 67–71.

57. "New Plantation at Cape Florida, Carolina," unpublished manuscript, box 1378.3, "Articles, Notes, Interview," Eric Norden Collection. See also *Pennsylvania Gazette*, 29 April 1731.

58. John Brickell, *The Natural History of North Carolina*, 4–8, 14.

59. "An Answer to the several Queries sent by the Lords Commissioners for Trade and Plantations," *CRNC*, vol. 6, 608, 611.

60. Justin Williams, "English Mercantilism and Carolina Naval Stores, 1750–1776," *Journal of Southern History* 1, 169–185.

61. "Concerning Wild Cattle," Laws of North Carolina, 1673, Act IV, *The State Records of North Carolina* (*SRNC*), Walter Clark, ed., vol. 25, 504–505.

62. "An Act to Prevent the Unreasonable Distruction of Fish," 1 December 1766, *SRNC*, vol. 25, 504–505.

63. Gabriel Johnston to the Board of Trade, 15 October 1736, *CRNC*, vol. 4, 176.

64. Henry McCulloh to Secretary of Board of Trade, January 1738, *CRNC*, vol. 4, 285.

65. Josiah Martin, "A Proclamation," 26 May 1772, *CNRC*, vol. 9, 294–295.

66. William Palmer to Josiah Martin, 26 July 1774, *CRNC*, vol. 9, 1020; William Rutherford to Josiah Martin, 26 July 1774, *CRNC*, vol. 9, 1021.

67. "Militia Returns, 1754–1755," *SRNC*, vol. 22, 314.

68. "An Act to Prevent burning the Woods," Laws of North Carolina, 1777, Chapter XXV, *SRNC*, vol. 24, 134.

69. "An Act to Prevent the Pernicious Practice of Hunting with a Gun in the Night by Fire Light," Laws of North Carolina, 1774, Chapter VI, *SRNC*, vol. 23, 955.

70. "An additional Act to an Act, intitulated an Act, to prevent killing Deer at Unreasonable Times, and for putting a stop to many Abuses committed by White Persons, under Pretence of hunting," Laws of North Carolina, 1745, Chapter III, *SRNC*, vol. 23, 218–219.

71. "An Act to amend an Act, entitled, an Act to prevent killing Deer at Unseasonable Times, and for putting a stop to many Abuses committed by White Persons, under Pretence of hunting," Laws of North Carolina, 1766, Chapter XIII, *SRNC*, vol. 23, 775–776.

72. Gabriel Johnston, "Report on the Spanish Wrecks of 1750," *CRNC*, vol. 4, 1301.

73. "Virginia: Journal of the Council," 11 March 1719, *CRNC*, vol. 2, 327.

74. Letter to Lord Cartwright, 14 February 1719, *CRNC*, vol. 2, 324–327. For North Carolina's bad reputation, see also *William Byrd's Histories of the Dividing Line Betwixt Virginia and North Carolina*, William K. Boyd, ed., 58; and Paul D. Escott, *Many Excellent People: Power and Privilege in North Carolina, 1850–1900*, xv: "In the 1680s Governor Culpeper of Virginia dismissed North Carolina as 'always . . . the sinke of America, the Refuge of our Renegadoes.'"

75. "An Act to encourage Persons to settle in the Town of Brunswick on the Southwest Side of Cape Fear River," Laws of North Carolina, 1745, Chapter XII, *SRNC*, vol. 23, 239–243; Council Proceedings, 1 October 1748, Board of Trade Journals, *CRNC*, vol. 4, 891–892; introduction to *SRNC*, vol. 22, iii–iv; "The Publick Dr. to Joseph Blake," *SRNC*, vol. 22, 276.

76. Laws of North Carolina, 1760, Chapter VIII, *SRNC*, vol. 23, 535–537.

77. Gwen Causey, *Great Is Thy Faithfulness: A Glorious Century, A Centennial History, 1899–1999*, 77, writes that in 1762 pastor Ezekiel Hunter from Onslow County baptized a few members at Lockwoods Folly and helped them build a log meeting house on the river bank, starting what has survived to this day as Mount Pisgah Baptist Church.

78. Catharine Phillips, *Memoirs of the Life of Catharine Phillips*, 80–82. See also David Cecelski, *A Historian's Coast*, 12–14.

79. Richard J. Hooker, ed., *The Carolina Backcountry on the Eve of the Revolution: The Journal and Other Writings of Charles Woodmason, Anglican Itinerant*, 80–81.

80. Nicholas Christian to the Society for Propagating the Gospel, 27 July 1774, *CRNC*, vol. 9, 1022–1023.

81. Peter H. Wood, "The Changing Population of the Colonial South: An Overview by Race and Region, 1685–1790," in *Powhatan's Mantle: Indians in the Colonial Southeast*, Wood, Gregory A. Waselkov, and M. Thomas Hatley, eds., 35–103, especially 43–46, 49. Wood estimates that by 1790 fewer than 400 Indians remained in North Carolina east of the mountains. See also Timothy Silver, *A New Face on the Countryside*, 69; Douglass LeTell Rights, *The American Indian in North Carolina*, 2nd ed., 122, 144–149; and Thomas E. Ross, *American Indians in North Carolina: Geographic Interpretations*. The lineage of the Lumbee is shrouded in myth: part white, part Indian, some say with a touch of black, they are rumored to be descendants of Roanoke's first settlers. They gained federal status as a tribe only in the twentieth century. See Gerald Sider, *Living Indian Histories: The Lumbee and Tuscarora People in North Carolina*.

82. Unsigned letter to the *North Carolina Gazette*, 5 December 1777, *SRNC*, vol. 11, 812–813.

83. For an insightful interpretation of colonists' frightful attitudes toward wolves, as well as the outlandish tales and vicious extermination campaigns they inspired, see Jon T. Coleman, *Vicious: Wolves and Men in America*, especially 37–65.

84. "An Act to Amend and Act, intituled 'An Act for Destroying Vermin in this Province,'" Laws of North Carolina, 1754, *CRNC*, vol. 1, 312.

85. A classic overview of a similar frontier economy is Peter H. Wood, *Black Majority: Negroes in Colonial South Carolina from 1670 through the Stono Rebellion*, 28–34, 95–130; see also Marvin L. Michael Kay and Lorin Lee Cary, *Slavery in North Carolina, 1748–1775*, 15–16, 44–45.

86. Bradford Wood has superbly and exhaustively described the colonial development of the Lower Cape Fear in *This Remote Part of the World: Regional Formation in Lower Cape Fear, 1725–1775*; see also A. Roger Ekirch, *"Poor Carolina": Politics and Society in Colonial North Carolina, 1729–1776*, 19–47, 74–75, 83–84.

87. Richard Dunn, *Sugar and Slaves: The Rise of the Planter Class in the English West Indies, 1624–1713*, 59–67, 73, 191–201, 319; Philip D. Curtin, *The Rise and Fall of the Plantation Complex: Essays in Atlantic History*, 2nd ed., 73–85; also Roderick A. McDonald, *The Economy and Material Culture of Slaves: Goods and Chattels on the Sugar Plantations of Jamaica and Louisiana*; and Stuart B. Schwartz, *Sugar Plantations in the Formation of Brazilian Society : Bahia, 1550–1835*.

88. James Clifton, "Golden Grains of White: Rice Planting on the Lower Cape Fear," *North Carolina Historical Review* 50, 365–393; James L. Sprunt, *The Story of Orton Plantation, with References to Other Places of Interest on the Cape Fear River, as Gathered from Various Histories and Documents*, 4; Joyce Chaplin, "Tidal Rice Cultivation and the Prob-

lem of Slavery in South Carolina and Georgia, 1760–1815," *William and Mary Quarterly* 49, 29–61, and *An Anxious Pursuit: Agricultural Innovation and Modernity in the Lower South, 1730–1815*, chapter 7, especially 266–267; Mart Stewart, *"What Nature Suffers to Groe": Life, Labor, and Landscape on the Georgia Coast, 1680–1920*, 98–114, 126–128; William Dusinberre, *Them Dark Days: Slavery in the American Rice Swamps*, 5–12, 48–56; and Wood, *Black Majority*, 35–62.

89. Wood, *Black Majority*, 112–113; Percival Perry, "The Naval-Stores Industry in the Old South, 1790–1860," *Journal of Southern History* 34, 509–526; Carroll B. Butler, *Treasures of the Longleaf Pines Naval Stores*; Robert B. Outland, *Tapping the Pines: The Naval Stores Industry in the American South*. Descriptions of naval stores production also abound in the travel literature; see Hodgson, *Letters from North America*, 39; Amelia M. Murray, *Letters from the United States, Cuba, and Canada*, 194–195.

90. David Cecelski, *The Waterman's Song: Slavery and Freedom in Maritime North Carolina*, 25–56; Outland, *Tapping the Pines*, 60–97.

91. The longleaf pine belt, often deemed unfit for large-scale agricultural development, constituted somewhat of a barrier between coastal plantation areas and the farmlands of the upcountry. As a result, North Carolina's coastal communities had stronger maritime ties to the Caribbean and North American port cities than it had inland ties to towns like Charlotte or Raleigh. U. B. Phillips, *A History of Transportation in the Eastern Cotton Belt to 1860*, pointed out that settlement of the tidewater forest region resulted from north-south coastal migration; settlers to the upcountry beyond the longleaf pine belt tended to come south from central Pennsylvania rather than west from the eastern coast. These diverging migration patterns combined with the perceived barrenness of the savannas to hamper the development of east-west transportation routes that could have stimulated trade between upcountry and coast.

92. Franklin W. Knight and Peggy K. Liss, eds., *Atlantic Port Cities: Economy, Culture, and Society in the Atlantic World, 1650–1850*, 2–3.

93. Hogg & Clayton to Alexander Duncan, 10 July 1763; Hogg & Clayton to Richard Quince, 10 July 1763; Hogg & Clayton to William Purviance, 10 July 1763; Hogg & Clayton to Alexander Duncan, 12 July 1763, Hogg & Clayton Letterbook.

94. Alan D. Watson, *Wilmington: Port of North Carolina*, 3–23.

95. John Bartram, *Diary of a Journey through the Carolinas, Georgia, and Florida from July 1, 1765, to April 10, 1766*, 15–18.

Chapter 2. "Turp^{t} Is the Word!"

1. William Bartram, *Travels through North and South Carolina, Georgia, East and West Florida*, 471–475; Thomas Slaughter, *The Natures of John and William Bartram*, 176.

2. Gordon S. Wood, *The Radicalism of the American Revolution*, especially 325–347.

3. See Edward J. Cashin, *William Bartram and the American Revolution on the Southern Frontier*.

4. Richard White, *The Middle Ground: Indians, Empires, and Republics in the Great Lakes Region, 1650–1815*, explains the ways in which shifting alliances among Indians

and between Indians and whites shaped the course of revolutionary conflict—the defeat of the British essentially marked the end of the Indians' ability to play Europeans and Americans against each other.

5. "Life and Travel in the Southern States," *Great Republic Monthly* 1, 80–84.

6. Hodgson, *Letters from North America*, 39–40.

7. Thomas C. Cochran, *Frontiers of Change: Early Industrial Development in the United States and Britain*, 5–16.

8. This influential term was established as a paradigm by Charles Grier Sellers, *The Market Revolution: Jacksonian America, 1815–1846*, 19. Note, though, that Sellers maintains that "Capitalist transformation *invaded* the southern and western interior and postwar boom galvanized the market culture into market revolution" (my italics).

9. See William Gordon to the Secretary, 13 May 1709, *CRNC*, vol. 1, 715; Records of General Court, 31 March 1713, *CRNC*, vol. 2, 81. Further showing the importance of naval stores are the appropriation by North Carolinians of the derogatory term "tar heels" as a nickname, as well as such colorful North Carolina sayings as "slower than tar in cold moonshine" and "as black as tar." Newman Ivey White et al., eds., *The Frank Brown Collection of North Carolina Folklore*, vol. 1, 484–485.

10. Robert W. Brown to Oliver H. Jones, 2 September 1835, folder 2, Oliver H. Jones Papers.

11. The evidence for this shift is admittedly sketchy; even noted naval stores historian Percival Perry failed to provide much concrete evidence for his claim that during "earlier days naval stores had been produced principally by the small farmers."

12. One exception was Josiah Swift, who in 1804 described how "the slaves of General Smith commenced the burning of lime in pens, called kilns, formed of sapling pines formed in squares containing from one thousand to one thousand two hundred bushels of oyster shells." Kemp P. Battle, *Letters and Documents, Relating to the Early History of the Lower Cape Fear*, 96.

13. Winslow C. Watson, ed., *Men and Times of the Revolution, Or: Memoirs of Elkanah Watson, Including His Journals of Travels in Europe and America from the Year 1777 to 1842 and His Correspondence with Public Men, and Reminiscences and Incidents of the American Revolution*, 2nd ed., 47, 50; Antonio Pace, trans. and ed., *Luigi Castiglioni's Viaggio: Travels in the United States of North America, 1785–87*, 118.

14. Robert Brown to Oliver Jones, 19 July 1833, folder 1, Oliver H. Jones Papers.

15. Robert W. Brown, "The Trade of Wilmington, 1843," in James Sprunt, *Chronicles of the Cape Fear River, 1660–1916* 2nd ed., 506–511.

16. Murray, *Letters from the United States, Cuba, and Canada*, 195. This particular letter is dated 12 January 1855.

17. See Jeffrey J. Crow, "Slave Rebelliousness and Social Conflict in North Carolina, 1775 to 1802," *William and Mary Quarterly*, 3rd Series, vol. 37, 79–102.

18. First U.S. Census, 1790; Second U.S. Census, 1800. In 1790 slaves comprised 40% of the population; in 1800 that number had fallen to 27%. The number of "other free" rose from 338 to 647.

19. F. P. Porcher, "Resources of the Southern Fields and Forests," *De Bow's Review*, August 1861, 105–131.

20. "A short account of Lumberton, in this State," *Hall's Wilmington Gazette*, 30 August 1798.

21. Watson, *Wilmington: Port of North Carolina*, 36. An earlier project to improve local transportation facilities entailed the incorporation in 1805 of the Long Creek Company, which was to build a canal from Lockwoods Folly to the Elizabeth River in Brunswick County.

22. Robert Brown to Oliver Jones, 2 September 1835, folder 2, Oliver H. Jones Papers.

23. Watson, *Wilmington: Port of North Carolina*, 7.

24. *Carolina Observer*, 22 August 1816.

25. *Raleigh Register and North Carolina Gazette*, 9 May 1823, newspaper clipping in box 6, file 4, Bob High Papers.

26. *Wilmington Herald*, 1 September 1826, newspaper clipping in box 6, file 4, Bob High Papers.

27. A claim made by the Wilmington *People's Press*, reprinted in *Carolina Observer*, 22 January 1833.

28. Cronon, *Nature's Metropolis: Chicago and the Great West*, 263–309.

29. Clifton Crawford, *Bladen County History*, 7–8, 24–25; the Second U.S. Census shows that the town had 65 inhabitants in 1800.

30. "A short account of Lumberton, in this State," *Hall's Wilmington Gazette*, 30 August 1798.

31. On western expansion east of the Mississippi, see John Mack Faragher, *Sugar Creek: Life on the Illinois Prairie*; Andrew Cayton, *The Frontier Republic: Ideology and Politics in the Ohio Country, 1780–1825*; Steve Aron, *How the West Was Lost: The Transformation of Kentucky from Daniel Boone to Henry Clay*; Alan Taylor, *Liberty Men and Great Proprietors: The Revolutionary Settlement on the Maine Frontier, 1760–1820*.

32. Third U.S. Census, 1810; Fourth U.S. Census, 1820: between 1810 and 1820, Brunswick County's population rose from 4,778 to 5,480; Columbus' population grew from 3,022 to 3,912; Bladen's from 5,671 to 7,276; Robeson's from 7,528 to 8,204.

33. First U.S. Census, 1790; Second U.S. Census, 1800; Third U.S. Census, 1810; Fourth U.S. Census, 1820.

34. The fact that Wilmington merchants toward the end of the eighteenth century were likely to receive large rather than small shipments did not necessarily reflect the continued dominance of large naval stores producers and their slaves over smaller producers. In what appear to be standard transactions, the mercantile business of John Huske and James Hogg made fourteen purchases of naval stores between April of 1783 and November of 1789. They paid their suppliers an average amount of about £60, a substantial sum for what must have been a substantial amount of produce. Only two transactions involved small amounts: one for £2 s16, and one for £3 s15, the next lowest being £27, Ledgers, John Huske and James Hogg Papers. This pattern may seem

to suggest that large producers were much more likely than small farmers to ship their forest produce to Wilmington. Given the nature of rural economies in the early republic, especially with their growing dependence on towns, it is far more likely that the larger suppliers purchased naval stores from small producers in their neighborhoods or acted as their agents in Wilmington. Moreover, because the production of pitch, rosin and spirits was complicated and troublesome and yields were generally low, many small and large producers limited themselves to the production and shipment of crude turpentine. As a result, there was a degree of parity between them as both produced for a largely singular market.

35. Lee Fleming to John McRae, 7 August 1848, Hugh MacRae Papers.

36. *Carolina Observer*, 11 January 1832.

37. During its 1833–1834 session, the North Carolina legislature passed acts to incorporate the North Carolina Centre and Seaport Rail Road Company, the Wilmington and Raleigh Rail Road Company, the Lumber River and Cape Fear Rail Road Company, and the Whitesville, Waccamaw and Cape Fear Canal and Rail Road Company. *Carolina Observer*, 21 January 1834.

38. Lee Fleming to John McRae, 13 April 1849, Hugh MacRae Papers.

39. Perry, "The Naval-Stores Industry in the Old South."

40. "Turpentine and Its Uses, *Southern Cultivator* 4 (1846), 104–105.

41. Sixth U.S. Census, 1840; Eighth U.S. Census, 1860.

42. Sixth U.S. Census, 1840; Seventh U.S. Census, 1850; Eighth U.S. Census, 1860; Jane Turner Censer, "Southwestern Migration among North Carolina Planter Families: 'The Disposition to Emigrate,'" *Journal of Southern History* 57, 407–426.

43. Eighth U.S. Census, 1860.

44. "The Turpentine Business," *Carolina Observer*, reprinted in *Southern Cultivator* 11 (1853), 209.

45. Sixth U.S. Census, 1840; Seventh U.S. Census, 1850; Eighth U.S. Census, 1860: Robeson County appears to have been a more typically "southern" county where slaveholders moved their slaves in order to grow cotton, rice, and tobacco.

46. Perry, "Naval Stores Industry of the Old South."

47. Reprinted in *Carolina Observer*, 10 April 1839.

48. *Carolina Observer*, 1 May 1839, 8 May 1839, 15 May 1839, 22 May 1839, 29 May 1839, and 5 June 1839.

49. "The Long Leafed Pine," *Wilmington Journal*, reprinted in *Southern Cultivator* 19 (1861), 243.

50. Murray, *Letters from the United States, Cuba, and Canada*, 197. This particular letter, from Charleston, is dated 7 January 1855.

51. Robert Potter Diary, entries for 21 August 1806, 26 August 1813, 4 September 1815, and 28 September 1822, Potter and Platt Family Papers. Potter reported an earthquake on 7 February 1812. See also *Carolina Observer*, 30 August 1827, 3 September 1829, 8 July 1830, 22 July 1830, and 26 August 1830; *Raleigh Minerva*, 4 October 1816, newspaper clipping, box 6, file 3, Bob High Papers.

52. "Public Instrument of Declaration and Protest," 11 October 1842, "Notarial Office's Repair Order," 1 October 1842, "Bill," 10 October 1842, "Auction Bill," 10 October 1842, Henry L. Adams Papers.

53. Enoch Rust to William Walter & Company, 4 January 1810, box XIII-D, William Walter and Company Papers.

54. Enoch Rust to William Walter & Company, 15 May 1811, box XIII-D, William Walter and Company Papers.

55. *Carolina Observer*, 6 September 1837.

56. "An Act for Destroying Wolves, Wildcats, Panthers, Bears, Crows and Squirrels in the Several Counties Therein Mentioned," Laws of North Carolina, 1785, *SRNC*, vol. 24, 749–750.

57. "An Act to amend an Act, intituled, An Act to Prevent Burning the Woods," Laws of North Carolina, 1782, Chapter XXIX, *SRNC*, vol. 24, 460; "An Act to Keep Open Cape-Fear, Wharie, Rockey River of Pee-Dee, the Catawba, Deep River, the Yadkin River, Dan, Haw River and Rockey River, for the Passage of Fish up the same," Laws of North Carolina, 1786, *SRNC*, vol. 24, 849–850; "An Act to prevent the several species of hunting therein mentioned," Laws of North Carolina, 1784, Chapter XXXIII, *SRNC*, vol. 24, 595–597.

58. *Carolina Observer*, 23 December 1824.

59. Thomas Jones, *The Experience of Thomas H. Jones, Who Was a Slave for Forty-three Years*, 7.

60. *A Narrative of the Adventures and Escape of Moses Roper, from American Slavery*, 48–49.

61. Raleigh *Minerva*, 23 July 1824, newspaper clipping, box 6, file 4, Bob High Papers.

62. Cecelski, *The Waterman's Song*, 128–129.

63. Richard M. Lewis and 21 others to Thomas Bragg, 25 August 1856, Thomas Bragg, Governors Letter Books, 514–515.

64. Thomas David Smith McDowell to Gillespie & Robeson, 27 January 1845, Thomas David Smith McDowell Papers.

65. "An Additional Act to Amend the Several Acts for Regulating the Town of Wilmington, and to Regulate and Restrain the Conduct of Slaves and Others in the Said Town, and in the Towns of Washington, Edenton and Fayetteville," Laws of North Carolina, 1785, Chapter VI, *SRNC*, vol. 24, 725–729.

66. *Raleigh Minerva*, 23 June 1820, newspaper clipping, box 6, file 4, Bob High Papers.

67. W. McKee Evans, *To Die Game: The Story of the Lowry Band, Indian Guerrillas of Reconstruction*, is the classic account of this episode; Pocahontas Wight Edmunds, *Tales of the North Carolina Coast*, 99, referred to the Lowrie gang as "shrewd cunning half-breeds."

68. See Joyce Appleby, *Capitalism and a New Social Order: The Republican Vision of the 1790s*, 5. Appleby sees the Republican vision as liberating Americans from the Federalists'

clinging to Anglophile traditions of deference, aristocracy, and monarchical grandeur, as well as the social stratification they engendered. It seems, however, that the rapidity of frontier settlement and the capitalist development it facilitated greatly outran the expansionist policies of the Jeffersonian Republicans—hence the revival of the Hamiltonian vision for the American economy, which carried within itself the mechanisms for the containment of certain of its excesses and uncertainties by enterprising elites.

69. See Larry E. Tise, *The American Counterrevolution: A Retreat from Liberty, 1783–1800*, for elite maneuvers to prevent the revolutionaries from taking the spirit of liberty too far.

70. Whether we can classify the antebellum southern economy as "capitalist" has long attracted the attention of scholars, and the debate reflects the dominant understanding of capitalism as a form of economic freedom that liberated Americans from traditional socioeconomic restraints on their economic pursuits. In this interpretation, the South's continued use of slavery after the revolution constituted an obstacle to such freedom, and its retrenchment and spread during the antebellum era a sign of the region's conservatism in the face of progress. But if we see as equally—if not more—crucial to capitalism the emergence of an entrepreneurial class that takes the lead in organizing land, resources, and people in a way that brings profit and progress to those who subscribe to its worldview, there is no doubt that southern planters were at the forefront of developments that presaged the modern American economy.

71. In "throughput" economies, the main products are of little intrinsic value to the producers, especially when produced in large quantities; only their likely purchase and reinvestment toward the production of more valuable capital justify the initial allocation of capital and labor power toward resource extraction. Unlike unsold surpluses of corn, which could be used for liquor production, home consumption, local barter, seed, and livestock feed, large quantities of turpentine and tar were of little domestic use to the farmers and planters who produced them. Only their reinvestment in industries such as shipbuilding and rubber processing warranted their large-scale production. Without demand for naval stores, the ownership of these natural resources was close to worthless, and the expenditure of labor power to transform resources into capital an outright waste.

72. *Carolina Observer*, 8 May 1831.

73. Charles Sellers, *The Market Revolution*, identifies more than two, of course, and for the most part rightly so—the market revolution had immense cultural, religious, social, political, as well as economic implications for American life. See also Anne C. Rose, *Voices of the Marketplace: American Thought and Culture, 1830–1860*. Here I am concerned with the ways in which the transportation of extracted resources increased the demand for those resources, putting increased pressure on the natural resource base at the same time that the environmental limits on the enjoyment of those resources seemed to have been eliminated.

74. Williams & Murchison price list, 1877, John McKay Papers.

75. "North Carolina," *De Bow's Southern and Western Review* 1, 105–122.

76. "Turpentine," *Southern Cultivator* 4, 104–105; A Subscriber, "Making of Turpentine, Rosin, &c.—Information Wanted," *Southern Cultivator* 12, 119; "Turpentine, Rosin, Pitch and Tar," *Southern Cultivator* 12, 147–148; "Turpentine Business in Florida," *Apalachicola Advertiser*, reprinted in *Southern Cultivator* 12, 160; "Turpentine Product of the South," *Southern Cultivator* 13, 79; "Turpentine: Hints for Those about to Engage in Its Manufacture," *Southern Cultivator* 13, 268–270; S. B. Buckley, "Long Leafed Pine (Pinus Palustris)," *Southern Cultivator* 25, 282; Tar-Heel, "The Long Leaf Pine," *Southern Cultivator* 25, 366.

77. Herbert Anthony Kellar, ed., *Solon Robinson, Pioneer and Agriculturist: Selected Writings*, vol. 2, 219–223. Richard Grist owed $20 for four years' subscription to *Niles' Register*: Bill for Subscription, 27 November 1832, box 2, folder 3, James Redding Grist Papers.

78. *Raleigh Register* and *North Carolina Gazette*, 25 April 1823, newspaper clipping, box 6, file 4, Bob High Papers.

79. "Notes on the Long-Leafed Pine" and John MacLeod, "The Tar and Turpentine Business of North Carolina," *Farmers' Library* 2, 12–19; "North Carolina," *De Bow's Southern and Western Review* 2 (new series), 105–122; "Product of Turpentine at the South," *De Bow's Southern and Western Review* 2, 303–305; "Turpentine: Hints for Those about to Engage in Its Manufacture," *De Bow's Review* 6, 486–489; Clingman, "North Carolina: Her Wealth, Resources, and History," *De Bow's Review* 9, 664–679; "The Pine Trees of Lower North Carolina and Virginia," *De Bow's Review* 13, 325–327; "The Pine Forests of the South," *De Bow's Review* 13, 72–75; "The Yellow Pine Forest of the South," *De Bow's Review* 8, 535–536.

80. Gavin Wright, *The Political Economy of the Cotton South: Households, Markets, and Wealth in the Nineteenth Century*, 45, 116–117.

81. Before Robert Outland's *Tapping the Pines* set them straight, historians of American slavery often overlooked this important allocation of slave labor toward industrializing the forest economy of the American South, in part because census reports do not describe the kind of auxiliary slave tasks that turpentine extraction often constituted. Having taken their cue from Wood's *Black Majority*, they commonly understood naval stores production as one aspect of a fluid system of forced labor on the colonial frontier in which slaves performed multiple tasks and often worked side by side with their owners. With the maturation of the southern economy and the hardening of racial lines between master and slave, they argue, slavery became a more rigid system of labor exploitation in which slaves performed more specialized tasks, often in gangs. The extraction of forest products as one of several plantation tasks of equal importance fell by the wayside as the varied nature of labor that had characterized colonial slavery had diminished by the antebellum period. Most of the slave's interaction with the forest was confined within a culture of resistance—slaves hunted and fished to supplement their plantation diets, retreated to the woods for unauthorized religious worship, or hid in the swamps when on the run.

82. "Turpentine," *Southern Cultivator* (1846); Outland, "Slavery, Work, and the Ge-

ography of the North Carolina Naval Stores Industry, 1835–1860," *Journal of Southern History* 62, 27–56; Robert S. Starobin, "The Economics of Industrial Slavery in the Old South," in Paul Finkelman, ed., *Economics, Industrialization, Urbanization, and Slavery*, 483–526.

83. Sixth U.S. Census, 1840; Eighth U.S. Census, 1860. In 1840, when Bladen, Brunswick, Columbus, and Robeson produced a combined total of 25,414 barrels of tar, pitch, turpentine, and rosin, only Columbus County had 123 "men employed" in the forest products industry, which included lumber, ginseng, and skins and furs. In 1860 Bladen County had 226 establishments that together employed 318 men in the production of crude turpentine and 260 turpentine distilleries that employed 53 men; Brunswick's 41 turpentine orchards employed 238 men and its 13 distilleries employed 34 men; Columbus County had 26 distilleries with 519 men in their employ; Robeson's 11 distilleries employed 155 men. In comparison, Cumberland County's naval stores production was 1,794 barrels with only 4 "men employed," while New Hanover County, claiming 22 sawmills and 4 lumber yards, produced the following with 334 men employed: lumber, $14,178; naval stores, 35,951 barrels; skins and furs, $10. With Wilmington as its main city, New Hanover was much more likely than the more rural counties to have a free white labor force working in the forest industry, but obviously the industry employed both free white and forced black labor during the antebellum period. Although slavery was a central aspect of turpentine extraction, it was not as singularly fundamental to the naval stores industry as it was to cotton culture. After the Civil War, the somewhat obfuscated relationship between slave labor and naval stores manifested itself in different responses to the perceived effects emancipation had on the southern economy. A North Carolinian who signed his letter to the *Southern Cultivator* with "Tar-Heel" attributed a decline in turpentine exports from Wilmington to the fact that "free negroes will not work." Yet D. R. Eldred, "Turpentine," *Southern Cultivator*, writing from Texas to the same journal, expressed the opinion that "free negroes and cotton will not pay" and considered switching to turpentine production exactly because it appeared to him that the latter did not cause the kinds of conflicts over land and labor that cotton engendered across the South. Whereas southern legislatures after the Civil War went to great lengths to keep their former slaves tied to cotton plantations, the turpentine camps more regularly employed men of both races. Allowing a degree of racial fluidity and economic independence before the Civil War, the naval stores industry proved more easily adaptable to postwar conditions than the cultivation of the most important agricultural crops.

84. James Battle Avirett, *The Old Plantation: How We Lived in Great House and Cabin before the War*, 70.

85. Rosser Howard Taylor, "Slaveholding in North Carolina: An Economic View," in *James Sprunt Historical Publications* 18, 38–40.

86. This information comes from a Mercer family history written by Mattie Taylor Gill, based on oral history, in the Mercer Family Papers.

87. James Grist to Allen Grist, 4 February 1851, box 2, folder 5, James Redding Grist Papers.

88. James Grist to Allen Grist, 16 July 1852, box 2, folder 5, James Redding Grist Papers.

89. James Grist to Allen Grist, 4 February 1851, box 2, folder 5, James Redding Grist Papers.

90. James Grist to Mr. Perry, 9 September 1850, box 2, folder 5, James Redding Grist Papers.

91. Allen Grist to James Grist, 23 September 1850, box 2, folder 5, James Redding Grist Papers.

92. James Grist to Allen Grist, 17 September 1852, box 2, folder 5, James Redding Grist Papers.

93. S. J. Norrell to James Grist, 18 August 1853, box 3, folder 1, James Redding Grist Papers.

94. E. Patterson to James Grist, 3 August 1856, box 3, folder 2, James Redding Grist Papers.

95. Ben Grist to James Grist, 19 May 1852, and John T. Council to James Grist, 20 September 1857, box 2, folder 5, James Redding Grist Papers.

96. Alexander McRae to [illegible] McRae, 28 May 1848, Hugh MacRae Papers.

97. On a comparable analysis of Midwestern boosters, see Don H. Doyle, *Social Order of a Frontier Community: Jacksonville, Illinois, 1825–1870*; also Cronon, *Nature's Metropolis*, 31–41; and Charles Nelson Glaab, *Kansas City and the Railroads: Community Policy in the Growth of a Regional Metropolis*.

98. John McRae to Donald McRae, 18 February 1849, Hugh MacRae Papers.

99. "Circular," 1 August 1849, Hugh MacRae Papers.

100. See Christiaan Grootaert and Thierry van Bastelaer, "Social Capital: From Definition to Measurement," in *Understanding and Measuring Social Capital: A Multidisciplinary Tool for Practitioners*, 1–15, who define social capital as "the institutions, relationships, attitudes, and values that govern interactions among people and contribute to economic and social development"; Pierre Bourdieu, "The Forms of Capital," in J. G. Richardson, ed., *Handbook of Theory and Research for the Sociology of Education*, 214–258, defines social capital as "the aggregate of the actual or potential resources which are linked to possession of a durable network of more or less institutionalized relationships of mutual acquaintance and recognition"; Robert Putnam, *Bowling Alone: The Collapse and Revival of American Community*, 19, sees social capital as "the collective value of all 'social networks' and the inclinations that arise from these networks to do things for each other"; Nan Lin, *Social Capital: A Theory of Social Structure and Action*, 19, defines it as "investment in social relations with expected returns in the marketplace."

101. Peter Lawrence Cash to Allen Grist, 29 April 1820, Bank of Newbern to Richard Grist, 25 June 1819, and unknown to Richard Grist, 28 February 1824, box 2, folder 2, James Redding Grist Papers.

102. Matthew B. O'Neil to Richard Grist, 9 July 1827, O'Neil to Grist, 16 July 1827, William Durham to Grist, 30 July 1827, Tucker & Haly to Grist, 21 July 1827, O'Neil

to Grist, 2 August 1827, and O'Neil to Grist, 1827, box 2, folder 2, James Redding Grist Papers.

103. Shipping bill, 29 January 1861, box 8, folder 1; R. M. Blackwell & Company to A. & J. R. Grist, 4 September 1850, 6 November 1852, 21 June 1851, and 26 June 1851, box 2, folder 5, James Redding Grist Papers.

104. Bill of Lading, 20 December 1834, folder 1; Bill of Lading, 2 January 1835, Bill of Lading, 5 January 1835, Aaron Lazarus to Oliver Jones, 7 January 1835, A. Hewlett to Oliver Jones, 4 April 1835, Potter & Parsley to Oliver Jones, 14 October 1835, A. Hewlett to Oliver Jones, 12 March 1836, folder 2; Cyrus C. Stow to Oliver Jones, 21 December 1839, Samuel W. Jones to Oliver Jones, 5 March 1841, 19 March 1841, 19 April 1941, 16 December 1841, 18 March 1842, 4 May 1842, and 10 November 1842, Ness & Jones to Oliver Jones, 30 April 1843, folder 3, Oliver H. Jones Papers. Rachel Mordecai Lazarus to Ellen Mordecai, 26 April 1835, Aaron and Rachel Lazarus to Samuel Mordecai, 3 May 1835, box 5, folder 1, Jacob Mordecai Papers; Emily Bingham, *Mordecai: An Early American Family*, 131, 162–163, 172.

105. North Carolina, vol. 4, p. 105, R. G. Dun & Company Collection, Baker Library, Harvard Business School. This entry was made on 20 June 1856. On 30 December 1858, the agency reported that Council's company was "wealthy but in bad repute. Trade with this firm . . . as if you expected to be cheated."

106. North Carolina, vol. 7, p. 265F, R. G. Dun & Company Collection, Baker Library, Harvard Business School.

107. North Carolina, vol. 21, p. 105, R. G. Dun & Company Collection, Baker Library, Harvard Business School. T.H. Russell, a turpentine distiller in Shoe Heel Depot, Robeson County, had been forced to borrow money from his wealthy and well-connected father, David H. Russell.

108. North Carolina, vol. 4, p. 104, R. G. Dun & Company Collection, Baker Library, Harvard Business School.

109. North Carolina, vol. 21, p. 327, R. G. Dun & Company Collection, Baker Library, Harvard Business School. This entry was made on 2 October 1850.

110. North Carolina, vol. 4, handwritten note attached to first page of index to Brunswick County, R. G. Dun & Company Collection, Baker Library, Harvard Business School.

111. North Carolina, vol. 4, p. 107, R. G. Dun & Company Collection, Baker Library, Harvard Business School.

112. North Carolina, vol. 7, p. 265I, R. G. Dun & Company Collection, Baker Library, Harvard Business School. See Karen Halttunen, *Confidence Men and Painted Women: A Study of Middle-Class Culture in America, 1830–1870*, for the fear of charlatans sparked by the growth of cities where the absence of traditional forms of social control gave such villains a large, little policed playing field. Similar fears generated by the uncertainties of the market gave rise to monitors such as R. G. Dun & Company.

113. North Carolina, vol. 21, p. 333, R. G. Dun & Company Collection, Baker Library, Harvard Business School.

114. North Carolina, vol. 18, p. 102, R. G. Dun & Company Collection, Baker Library, Harvard Business School.

115. North Carolina, vol. 7, p. 265F, R. G. Dun & Company Collection, Baker Library, Harvard Business School.

116. North Carolina, vol. 4, p. 104S, R. G. Dun & Company Collection, Baker Library, Harvard Business School.

117. "Copies of Laws of North Carolina," box 1378.3, Eric Norden Collection.

118. Gavin Wright, *The Political Economy of the Cotton South: Households, Markets, and Wealth in the Nineteenth Century*, 45–46.

119. Josiah Collins to Governor Edward B. Dudley, 4 May 1837, James Kearney to Dudley, 11 August 1837, box 10, folder 6; Charles B. Shaw, "Expense Report," 11 October 1837, "Receipt" (undated), box 10, folder 7; Shaw, "Expense Account," April 1838 and April/May 1838, box 10, folder 10, State Board of Education Swamp Land Records.

120. "Survey Plats," box 12; Fourney George to Thomas Bragg, 17 February 1855 and 24 March 1855, box 12, folder 20, State Board of Education Swamp Land Records.

121. "Eric Norden Answers Pertinent Questions on State School Lands," Wilmington *Sunday Star News*, 28 January 1934; a manuscript copy of this article is located in box 1378.3, Eric Norden Collection. W. C. Kerr, "Report on the Swamp Lands" (typed manuscript, undated, marked "Read at Literary Board meeting May 25, 1866"), box 13, State Board of Education Swamp Land Records.

122. *Carolina Observer*, 24 August 1831, 9 December 1840.

123. Basil Hall, *Travels in North America, in the Years 1827 and 1828*.

124. *Carolina Observer*, 11 February 1830.

125. Clitherall Diaries, Series 1, Volume 5, 26–28.

126. *The Diary of Elizabeth Ellis Robeson: Bladen County, North Carolina from 1847 to 1866*, entries for 10 November 1849 and 1 December 1849.

127. Clitherall Diaries, Series 1, Volume 5, 50; Series 1, Volume 6, 21–22.

128. *Carolina Observer*, 14 June 1826 and 24 May 1827.

129. Clitherall Diaries, Series 1, Volume 4, 46, 48.

130. *Ibid.*, Series 1, Volume 5, 45.

131. *The Diary of Elizabeth Ellis Robeson*, entry for 19 October 1848; my italics.

Chapter 3. "Enterprise vs. The 'Grab Game'"

1. Lemuel Bannister to "Frankie," 24 June 1868, 28 June 1868, and 29 October 1868, Lemuel Bannister Papers.

2. Deed, box 1378.20, Eric Norden Collection. The company's date of incorporation is unknown, but it already sold land in 1852.

3. Lemuel Bannister, *Lake Waccamaw Lands, Columbus and Brunswick counties, North Carolina* (1903).

4. C. Vann Woodward, *The Origins of the New South, 1877–1913*. Most obviously built on Woodward's work is Edward L. Ayers, *The Promise of the New South: Life after Reconstruction*, especially vii. Woodward's hand, however, is felt just about everywhere in the

historiography of the postwar South. The subject of a book-length collection of essays and an autobiography, Woodward and his impact cannot be contained within a single endnote. See "C. Vann Woodward's *Origins of the New South, 1877–1913*: A Fifty-Year Retrospective," *Journal of Southern History* 67, 715–826.

5. The literature is extensive, but a good introduction to the racially charged land/labor dimensions of the transition from slavery to wage labor is Gavin Wright, *Old South, New South: Revolutions in the Southern Economy since the Civil War*. Historians have long debated whether change or continuity should dominate the story of the New South. Focusing especially on the socioeconomic attitudes and behavior of the region's enterprising factions, they continue to debate whether or not elite directives for postwar development constituted a capitalist break with a pre- or non-capitalist past. Were the men who directed the economic development of the postwar South the same enterprising class of men as those who lorded it over the antebellum plantation? Or did the rise to prominence of a new class of capitalists displace paternalism and patron/client ties as the dominant form of socioeconomic relations, transforming plantation and "moral" yeoman household economy alike? Did the market replace the household as the center of economic activity, or did it expand into a production/consumption complex that increasingly incorporated production units other than households? Did the legacy of plantation slavery impede the development of full-blown capitalism, or was it just one form of capitalism that gave way to another? Although the shadings and qualifications of their arguments are many and diverse, historians such as Stephen DeCanio and Robert Higgs, who see more continuity than change, argue that the exploitative behavior and attitudes of antebellum planters and merchants was no less capitalist than that of their postwar counterparts. See Stephen DeCanio, *Agriculture in the Postbellum South: The Economics of Production and Supply*, 4–15, 220–223; also Robert Higgs, *Competition and Coercion: Blacks in the American Economy, 1865–1914*. Historians such as Barbara Jeanne Fields and Joseph Reidy, who emphasize change over continuity, although equally diverse and multifarious in their approach and detail, argue that the shift from slavery to wage labor constituted too much of a watershed between a pre-capitalist and a capitalist economy to sustain claims of continuity. See Barbara Jeanne Fields, *Slavery and Freedom on the Middle Ground: Maryland in the Nineteenth Century*, 157–163, 248–249; also Joseph P. Reidy, *From Slavery to Agrarian Capitalism in the Cotton Plantation South: Central Georgia, 1800–1880*. Harold D. Woodman, "Economic Reconstruction and the Rise of the New South, 1865–1900" in John B. Boles and Evelyn Thomas Nolen, eds., *Interpreting Southern History: Historiographical Essays in Honor of Sanford W. Higginbotham*, 254–307, gives an extensive overview. On the plight of blacks, see Roger L. Ransom and Richard Sutch, *One Kind of Freedom: The Economic Consequences of Emancipation*; Ronald Lewis, *Black Coal Miners in America: Race, Class, and Community Conflict, 1780–1980*, 3–12 on slavery, 13–35 on convict labor, and 39–37 on the bleak conditions for black miners in the postbellum South; David A. Oshinsky, *Worse than Slavery: Parchman Farm and the Ordeal of Jim Crow Justice*; and Leon F. Litwack, *Trouble in Mind: Black Southerners in the Age of Jim Crow*, especially chapter 3. The issue of paternalism is as divisive for the New

South as it is for the Old, often because a cloying air of romanticism and benevolence attaches itself to the subject. Melton Alonza McLaurin, *Paternalism and Protest: Southern Cotton Mill Workers and Organized Labor, 1875–1905*, xiii–xiv, xviii, 56, 199–201, asserts that cotton mills were industrial plantations where laborers accepted the "entrepreneur as provider and protector," at least in part out of racial solidarity: "The worker evidently valued his racial integrity and his honor more than his means of obtaining a livelihood. At least, he feared the Negro more than he feared and resented management's nearly absolute power over him." David L. Carlton, *Mill and Town in South Carolina, 1880–1920*, 89–109, 243–248, 302, rightly accuses McLaurin of melodrama, rejects the characterization of mill villages as industrial plantations, nuances the racial issue by pointing out that rabid racism found few supporters among mill owners, and calls for careful use of the term "paternalism," the practice of which was always firmly in the interest of the firm. See also I.A. Newby, *Plain Folk in the New South: Social Change and Cultural Persistence, 1880–1915*, 258–286. If anything is clear from the broad sweep of southern historiography, it is that environmental destruction and exploitation of humans—paternalistic or not—remained as closely tied in the New South as they had been in the Old. See for example Albert Cowdrey, *This Land, This South: An Environmental History*, rev. ed., 103–124, for the unlimited exploitation of nature, and Jacqueline Jones, *The Dispossessed: America's Underclasses from the Civil War to the Present*, 73–201, who puts a suffering human face on that story.

6. Michael Williams, *Americans and Their Forests: A Historical Geography*, 238–288; Henry Clepper, *Professional Forestry in the United States*.

7. Cowdrey, *This Land, This South*, 90, 112–113; Ayers, *The Promise of the New South*, 123–131; Thomas D. Clark, *The Greening of the South*, 18, 20, 24, 25; Ronald L. Lewis, *Transforming the Appalachian Countryside*, 165–184.

8. U.S. Forest Service, *Timber Depletion, Lumber Prices, Lumber Exports, and Concentration of Ownership: Report on Senate Resolution 311*, 3rd ed., 13. Williams, *Americans and Their Forests*, argues in the same vein that between 1800 and 1940 the center of lumber production, in every instance urged by timber depletion, moved from the Northeast to the Midwest, thence to the South, and finally to the Pacific Northwest, everywhere leaving large cutover areas in its wake.

9. Alfred Chandler, *The Visible Hand: The Managerial Revolution in American Business*, 1–12: the rise of big business in the second half of the nineteenth century brought about a shift in business organization that resulted in the "visible hand" of management, not only superseding Adam Smith's "invisible hand"—the market—as the driving force behind business development, but also reaching far beyond the realm of business.

10. Philip Scranton, *Endless Novelty: Specialty Production and American Industrialization, 1865–1925*, 7, 9. According to Mary Klein, "Coming Full Circle: The Study of Big Business since 1950," *Enterprise & Society* 2, 425–460, giving human beings a place in the history of enterprise also motivated the work of Thomas Cochran.

11. AnnaLee Saxenian, *Regional Advantage: Culture and Competition in Silicon Valley and Route 128*, 2, 7: "The concept of an industrial system illuminates the historically

evolved relationship between the internal organization of firms and their connections to one another and to the social structures and institutions of their particular localities." To the three dimensions she identifies as constituting a system—local institutions and culture, industrial structure, and corporate organization—I think we must add the environment.

12. North Carolina Department of Conservation and Development, "Life of the Yellow Pine Lumber Industry in North Carolina (October 1924)," typed manuscript, box 3, Activities of the Board of Conservation and Development (ABCD) 1926–1935. The report stated that "only 25 of the 2333 mills reported have a production of 6 million feet, and over, per annum," with "only four operations producing over 15 million feet per annum." The unit production of the other 2,308 mills averaged 375,000 feet a year, while another 500 farm mills cut 5,000–75,000 feet annually.

13. *Lumberton Argus*, 12 February 1903; the *Southport Leader* asked on 22 December 1892, "Who Owns the Land?"

14. Charles S. Sargent, *Report on the Forests of North America*, 515–516. According to Sargent, Bladen County had 288,000,000 board feet of standing longleaf pine timber as of 31 May 1880; Brunswick had 141 million board feet; Columbus, 288 million board feet; Robeson, 864 million board feet.

15. Evangeline Walker Andrews and Charles McLean Andrews, eds., *Journal of a Lady of Quality*, 149.

16. *Carolina Observer*, 30 January 1823.

17. Firewood for heating and cooking even made its way to the market of smaller towns such as Fayetteville, whose vicinity abounded with timber. See *Carolina Observer*, 19 June 1823.

18. Robert W. Brown, "The Trade of Wilmington, 1843."

19. Receipts, 28 November 1853; Thomas H. Robison to J. W. Purdie, 17 December 1856; Account of J. W. Purdie to J. M. Robinson & Son, 1857; Account of J. W. Purdie to J. T. Council, 25 September 1858, James W. Purdie Papers.

20. Tax list, Alexander Elliot Papers. Elliot's total property consisted of nearly 5,390 acres worth $5,669.

21. "Memorandum of Wages Paid Hands," 22 June 1827, Alexander Elliot Papers.

22. Lists of chores performed by H. Elliot for A. Elliot, 1 August 1821 and 10 August 1821, Alexander Elliot Papers.

23. See O.G. Parsley's ad in the *Carolina Observer*, 1 November 1837: "To Saw Mill Owners—I will attend to the sale of Lumber in this place [Wilmington]. Having been engaged in the business for six or eight years past, I am under the impression my acquaintance with this and foreign markets is such as to enable me to the best advantage." Several such ads appeared in the paper during the following years.

24. Wood was either "merchantable" or "refuse"; when merchantable, it could be classified as "timber," "scantling," "lumber," "flooring," "boards," or "wide boards."

25. H. McKellar to Alexander Elliot, 5 April 1830, Alexander Elliot Papers.

26. Alexander Elliot to William Campbell, 17 July 1846, Alexander Elliot Papers.

27. North Carolina, vol. 18, pp. 76, 87, 89, R. G. Dun & Company Collection, Baker Library, Harvard Business School.

28. North Carolina, vol. 21, pp. 336, 351, 554, R. G. Dun & Company Collection, Baker Library, Harvard Business School.

29. North Carolina, vol. 4, p. 114, R. G. Dun & Company Collection, Baker Library, Harvard Business School.

30. *Whiteville News Reporter*, 15 October 1992, newspaper clipping, Bob High Papers. North Carolina, vol. 7, pp. 266I, 266P, 266U, R. G. Dun & Company Collection, Baker Library, Harvard Business School.

31. *North Carolina State Advertiser*, 28 October 1865.

32. One need only read Paul Wallace Gates, "Federal Land Policy in the South, 1866–1888," *Journal of Southern History* 6, 303–330, to get a grasp of the enormity of land-grabbing on the part of northern interests. Gates estimates that between 1877 and 1880 federal authorities sold 5.7 million acres in Florida, Alabama, Mississippi, Louisiana, and Arkansas. See also Williams, *Americans and Their Forests*, 242–243.

33. Columbus County Record of Deeds 1901–1902, vol. SS, 138, vol. TT, 128–131, 255–257, 572–573, vol. WW, 179; Columbus County Record of Deeds 1903–1904, vol. ZZ, 61–63.

34. *Lumberton Argus*, 30 October 1902, 4 December 1902, 29 October 1903, 4 November 1904, 18 November 1904, and 29 November 1904.

35. John K. Baldwin Diary, entries for 1 January 1867, 2 January 1867, 3 January 1867, 4 January 1867, 6 January 1867, 24 March 1867, 16 June 1868, 8 September 1868, 9 September 1868. "Index to Real Estate Conveyances, Bladen County, 1784–1961, Index to Deeds: Grantee."

36. William Harris Garland, a mill worker and contemporary of John Baldwin, constantly feared that his boss would be forced to stop his Wilmington mill because of a timber shortage: William Harris Garland to William Harris Garland, 5 May 1870, 13 May 1870, 20 May 1870, 14 July 1870, William Harris Garland Papers.

37. John K. Baldwin Diary, entries for 5–18 August 1867, 8 May 1869, 4 February 1871, 8 March 1871, 8 May.

38. George Nelson Hill, Diary and Account Book, 1869–1872, entries for 14 June 1869, 3–4 July 1869, 21 September 1869, 4 October 1869, 1 June 1870, 5 June 1870, 13 August 1871, 15 October 1871, 22 October 1871, and 7 December 1871.

39. U.S. Forest Service, *Timber Depletion, Lumber Prices, Lumber Exports, and Concentration of Timber Ownership*, 20, 21.

40. Gifford Pinchot and W. W. Ashe, "Timber Trees and Forests of North Carolina," *North Carolina Geological Survey Bulletin*, No.6, 39–41, 119–121, 122–123, 125–126, 131–133, 158, 161, 164, 166, 172, 174, 176, 178.

41. Williams, *Americans and Their Forests*, 254.

42. Ibid., 248–249. See also International Workers of the World, *The Lumber Industry and Its Workers*, 2nd ed., 15: "In 1889 William Baptist put a ground system in operation in a Louisiana swamp. It consisted of two large drums and an engine and

boiler mounted on a scow from which an endless cable passed out into the forest for a distance of one-half mile. This later developed into the modern "slack rope" system now used on pull boats. A third method, called the 'snaking system,' was a later development in the pine forests of the South." Also Howard A. Hanlon, *The Bull Hunchers: A Saga of Logging the Tidewater Low Country*, 238–258.

43. Pinchot and Ashe, "Timber Trees and Forests of North Carolina," 166.

44. Henry Disston & Sons, *The Saw in History*, 2nd ed., 45–47.

45. *Lumberton Argus*, 7 May 1903.

46. *Directory of American Sawmills*, 150–164, especially 151. Although the 1915 directory lists a much smaller overall number of establishments than the 1884 *Directory and Shipping Guide of Lumber Mills and Lumber Dealers*, the proportion of companies that cut and saw cypress and other swamp trees is larger.

47. Albert Bushnell Hart, *The Southern South*, 45.

48. *Whiteville News Reporter*, 10 January 1923.

49. See Robert S. Maxwell and Robert D. Baker, *Sawdust Empire*; Thad Sitton and James H Conrad, *Nameless Towns*; International Workers of the World, *The Lumber Industry and Its Workers*; Vernon H. Jensen, *Lumber and Labor*, especially 71–98; Jeffrey A. Drobney, *Lumbermen and Log Sawyers*; Steven Andrew Reich, "The Making of a Southern Sawmill World: Race, Class, and Rural Transformation in the Piney Woods of East Texas, 1830–1930."

50. "The Sharpest Saw," in W. C. Hendricks, ed., *Bundle of Troubles and Other Tarheel Tales*, 30–32. "John Henry of the Cape Fear," in the same collection, 37–51, tells the story of an incredibly strong steel driver working on a railroad who dies in a contest with a mechanized steamhammer. This, too, suggests the seemingly unstoppable rise of the machine in industrial work.

51. *Southport Leader*, 20 March 1902; *Lumberton Argus*, 23 August 1904 and 18 April 1905; *Whiteville News Reporter*, 13 May 1926 and 30 June 1927, *Red Springs Comet*, 15 December 1892.

52. *Southport Leader*, 25 September 1890, 9 October 1890, 22 January 1891, 19 February 1891, 26 February 1891, 5 May 1891, 25 June 1891, 9 July 1891, 16 July 1891, 28 July 1892, 6 July 1893, 13 July 1893, 27 July 1893, and 3 August 1893; also Bill Reaves, *Southport (Smithville): A Chronology*, vol. 2, 28, 35.

53. *Southport Herald*, 20 February 1908.

54. Reaves, *Southport (Smithville)*, 240, 260.

55. North Carolina, vol. 4, p. 104Z6, R. G. Dun & Company Collection, Baker Library, Harvard Business School.

56. Abstract of title, Waccamaw Lumber Company, box 1378.20, Eric Norden Collection. The bulk of the property at one point belonged to the New Jersey and North Carolina Land and Lumber Company, which sold 171,120 acres to a J. J. Wolfenden, who sold the same tract to the Waccamaw Land and Lumber Company on 1 May 1904. The *Southern Lumberman* reported on 10 May 1907 that the supplanting of the Waccamaw land and Lumber Company by the Waccamaw Lumber Company involved $1 million in

capital stock and 200,000 acres of timberland. On 25 February 1904, the *Lumberton Argus* identified Wolfenden as "the head of the syndicate which recently bought the famous Green Swamp Lands in Brunswick and Columbus counties."

57. Property deeds for Columbus, Bladen and Brunswick counties; Dun & Bradstreet's credit report on Waccamaw Lumber Corp. 1932, Pierce & Company Papers; "Map of Bolton District," Hugh McRae Papers.

58. *Southport Herald*, 20 February 1908.

59. "Mr. Baker and the Waccamaw Lumber Company," *Kin'Lin'* 3, 41–44. *Kin'Lin'* was the Local Heritage Publication of Hallsboro High School; this issue, archived at Duke University, is devoted to an oral history of the Waccamaw Lumber Company. See also a set of the plant's blueprints at the Forest History Society; and *Directory of American Sawmills*, 151.

60. Dun & Bradstreet's credit report on Waccamaw Lumber Corporation, 1932, Pierce & Company papers.

61. Charles Luke to Eric Norden, 21 May 1924, Eric Norden Collection.

62. *Southern Lumberman* 27, 25 May 1907 and 24 August 1907; *Wilmington Morning Star*, 4 January 1912.

63. The 1910 census for Waccamaw Township, where the Makatoka camp was located, identified six black households occupied by twelve loggers, one blacksmith, one railroad construction worker, and three black cooks that shared a cabin. In Bolton, some black citizens took advantage of this racial policy by opening their homes to boarders who worked at the saw mill.

64. *Kin'Lin'* 3; U.S. Census Bureau, 1910 Population Schedule, North Carolina, Brunswick County, and Fourteenth Census of Population 1920: North Carolina, vol. 9, Brunswick County.

65. Pinchot and Ashe, "Timber Trees and Forests of North Carolina," 166.

66. W. W. Ashe, "The Forests, Forest Lands, and Forest Products of Eastern North Carolina," *North Carolina Geological Survey Bulletin*, No. 5, 110.

67. Pinchot and Ashe, "Timber Trees and Forests of North Carolina," 158.

68. Pierce & Company to D. M. Flynn, 11 January 1913, box 106, folder 386, Pierce & Company. Papers.

69. Ledgers, box 78, folder 190, and file S-2776, Pierce & Company Papers.

70. D. M. Flynn to Pierce & Company, 17 April 1916; Pierce & Company to D. M. Flynn, 19 April 1916, 9 May 1916, and 1 January 1917, Pierce & Company Papers.

71. Ledger, pp. 150–155, box 78, folder 190, Pierce & Company Papers.

72. C. Van Leuven to Pierce & Company, 1 October 1915; Pierce & Company to CTDC, 5 November 1915, Pierce & Company Papers.

73. Agreement between Geo. W. Frink and Pierce & Company (by J. B. Wyche), 1 January 1923, Pierce & Company Papers.

74. Agreement between W.L. Burchett and Pierce & Company (by J. B. Wyche), 17 June 1922, Pierce & Company Papers.

75. J. B. Wyche to Louis F. Ward, 1 June 1928, series 2.8, box 237, folder 1749, Pierce & Company Papers.

76. Weekly Payrolls, "Total Lumber Bought," 12 September 1930 to 4 September 1931, Pierce & Company Papers.

77. Weekly Payrolls, "Lumber Bought," 12 September 1930 to 4 September 1931, Pierce & Company Papers.

78. Pierce & Company to D. M. Flynn, 29 March 1917 and 3 September 1917; log statements, 2 April 1917 and 4 August 1917, box 106, folder 136, Pierce & Company Papers.

79. Ledger, box 78, folder 190, Pierce & Company Papers. Several deeds came without such a specification, but all deeds specified at what height the trees should be cut (usually one or two feet above the ground).

80. Pierce & Company stuck around, but its repeated complaint that the timber it agreed to cut was "very small," as well as the trouble it experienced locating enough longleaf pine of sufficient quality, suggests that it increasingly harvested second-growth trees or timber that other companies had left standing on account of its small size. Ultimately, it got out of the lumber business because it became cheaper to have lumber from the west coast delivered to Hallsboro. Pierce & Company to D. M. Flynn, 14 April 1916 and 27 September 1916, box 106, folder 386; Pierce & Company to Pearsall Company, 22 May 1930 and 27 May 1930, box 109, folder 421; Lu Ann Jones, transcript of interview with Cyril James Wyche, 9, series 2.1.1, folder 1812, Pierce & Company Papers.

81. Weekly Payrolls, "Work," 19 September 1930 to 11 September 1931; Weekly Payrolls, "Lumber," 2 February 1925, 11 September 1925, 15 January 1926, 19 March 1926, 11 June 1926, 18 June 1926, and 30 June 1926, volume 194 b–c; 16 September 1927 to 4 November 1927, volume 194 b, Pierce & Company Papers.

82. Transcript of interview with Cyril James Wyche, 20, Series 2.1.1, folder 1812, Pierce & Company Papers. John K. Baldwin's Diary suggests that change indeed came slowly to this rural economy. Entries for 4 January 1867, 12 January 1867, 24 January 1867, 28 January 1867, 22 July 1867, and 21 August 1872 show that Baldwin on one occasion bartered coal for a saw, was sometimes paid money for his labor, and other times accepted items such "buck-gloves," "pants," "glasses of jelly," and cloth and thread as payment for his work. Yet he always made note of the monetary value of the items he received.

83. Thirteenth U.S. Census, 1910; Fourteenth U.S. Census 1920; Fifteenth U.S. Census, 1930; Fourteenth Census of Population, 1920, North Carolina, vol. 9, Brunswick County.

84. J. B. Wyche to D. M. Flynn, 9 May 1916, Pierce & Company Papers

85. Pierce & Company to E. B. Kearns, 23 January 1947, box 143, folder 654; ledger, 9, 219, box 78, folder 191; Wyche interview, 34, Pierce & Company Papers.

86. *Whiteville News Reporter*, 30 September 1926.

87. "Crop Lien" folder; Wyche interview, 12, Pierce & Company Papers.

88. Whereas the Waccamaw Lumber Company obtained its heavy machinery from Michigan and erected its own repair shop, Pierce & Company bought its equipment used or from regional manufacturers, relied on local mechanics to take care of repairs,

and forged ties with the area's other forest-related companies, all of whose survival rested on the interdependence of smaller businesses. Within the area, Pierce & Company cultivated business relations with the Empire Tie Company of Hallsboro; the Standard Supply Company of Wilmington, which supplied one of the company's boilers; the Pearsall Company, a fertilizer company of Wilmington; and men such as J. J. Carroll and J. W. Lashley of Acme, who sold the company, second-hand a tram-road, an edger, a small boiler, a sawmill engine and a complete sawmill. The Atlantic Coast Line Railroad purchased crossties from Pierce & Company, and, deferring to local expertise, asked the company to supply the Standard Veneer Company of Maine with information regarding the possibility of establishing a veneer plant in Hallsboro. On a larger regional level, Pierce & Company had business dealings with the Liddell Company, a Charlotte-based machine manufacturer; the Carolina Machinery Company of Sumter, South Carolina; the Carolina Refractories Company of Hartsville, South Carolina, which specialized in lumber industry furnaces; the North State Lumber Company of Greensboro, which purchased lumber—sawed to specification—for one of its buildings; and New Bern Ironworks. Because the reach of these companies was largely local and regional, they depended on each other's patronage, products, knowledge, and expertise for the development of the regional economy. Because most of them were, at least in principle, committed to staying put, their network also served to sustain the forest industry in an area where natural resources were increasingly less widely and less readily available. Lumbering enterprises such as Pierce & Company especially benefited from technological advances in the sawing and drying of lumber that helped it generate greater yields from less timber. See "Tables of Costs of Plants, Machinery, etc.," 28 July 1920, folder 190; ledger, 36–40; Correspondence with ACLRR, box 105, folder 367; Carolina Refractories Company to Pierce & Company, 14 April 1937, box 106, folder 381; dealings with North State Lumber Company, box 109, folder 419; dealings with Pearsall Company, box 109, folder 421; G. A. Cardwell to Pierce & Company, 11 January 1918, J. M. Wyche to Allen Quimby, 17 January 1918, box 105, folder 338, Pierce & Company Papers.

89. John Stover, *The Railroads of the South, 1865–1900: A Study in Finance and Control*, 264; *The Lumberman's Directory of Saw Mills, Shingle Mills and Other Wood Working Factories*, 139, 141; Rand McNally & Company's *Directory and Shipping Guide of Lumber Mills and Lumber Dealers in the United States and Canada* (1884), 324–343; Eugene McMillan Musselwhite, "The 'Iron Horse' Comes to Robeson: A County of a 'Rip Van Winkle State,'" typed manuscript, 6, 7, 8, 13, 18, 20, 22–23, 24–25; *Lumberton Argus*, 31 December 1903; Ann Courtney Ward Little, ed., *Columbus County, North Carolina: Recollections and Records*, 78–84 and 140–151; *Red Springs Citizen*, 9 September 1898.

90. North Carolina, vol. 18, p. 93, R. G. Dun & Company Collection, Baker Library, Harvard Business School; *The Southern Business Directory and General Commercial Advertiser*, vol. 1, 397, indicates the existence in 1854 of Chadbourn & Company in Wilmington; correspondence, James H. Chadbourn and Joy and Metcalf, 6 April 1859, 14 March 1860, 24 March 1860, 13 October 1860, and 29 October 1860, Joy and Metcalf Papers.

91. Glenn F. Strole, *Chadbourn and her Sunny South Colony: A Narrative History*, rev.

ed.; Ward Little, *Columbus County, North Carolina*, 205–207; Columbus County, Record of Incorporations, 1887–1938 shows that the Chadbourn Mill and Railway Company was incorporated on 29 January 1887 with $24,000 in capital stock. James H. Chadbourn, George Chadbourn, William H. Chadbourn, and James H. Chadbourn Jr., each held 60 $100 shares.

92. James H. Chadbourn Jr., "To Whom it May Concern," 28 August 1899, folder 13, Brown Family Papers; James Alderman and E. Turlington to E. Kidder & Sons, James H. Chadbourn & Company, Northrop & Cumming, Parsley & Wiggins, and Holville & Company, no date, folder 36.1.a, Samuel James Rowell Papers.

93. Charles B. Gault to H. L. Lyon, 17 January 1924, folder 2, Charles Beers Gault Papers.

94. Charles Oscar Beers to Thomas H. Wright, Esq., 20 March 1912, box 1, folder 1, Charles Beers Gault Papers.

95. "Minnie" to unknown, 10 April 1909, box 1, folder 1, Charles Beers Gault papers; my italics.

96. Town of Whiteville, "Resolution," 17 May 1943, box 11, folder 2, Hugh MacRae Papers.

97. Katie Rowell to unknown, 6 December 1892; S. J. Rowell to Minnie Rowell, 5 June 1894; note, dated 3 September 1908, on back of letter from S. J. Rowell to Minnie Rowell, 7 November 1908, folder 36.1.a; receipt for sale of 213 Gum Logs, 21,645 feet. at $4.50 ($97.40) to Wilmington Cooperage Company, folder 36.2.b; diary entries for 18 February 1907 to 20 March 1907, folder 36.1.c; Pierce's Memorandum Account Book; notebook, no cover, entry for 11 December 1911, Samuel James Rowell Papers; my italics.

98. *Southport Standard*, 12 July 1900.

99. Articles from the *Progressive Farmer*: "Woodman Spare the Tree" and "Take Care of the Forests," 7 April 1886; "Working the Timber," 16 June 1886; "The Swamp Willow—How to Destroy It," "A Timber Speculation," "Timber Lands Should Be Properly Assessed," "Some Snakes You Know," 13 September 1906; for an example of a Salem Iron Works ad, see 17 September 1895; my italics.

100. *Southport Herald*, 19 April 1905.

101. *Southport Leader*, 11 April 1892; *Southport Herald*, 30 April 1908.

102. Note marked "Cavenaugh," 3 September 1901, Charles Beers Gault Papers.

103. "Notice," December 6, 1872, John McKay Papers; North Carolina, vol. 21, p. 335, R. G. Dun & Company Collection, Baker Library, Harvard Business School.

104. C. R. Humphreys to Francis Gault, 21 February 1911, Charles Beers Gault Papers. See also J. Billard Bellamy to James Sprunt, 20 October 1906, box 2, folder 4, Alexander Sprunt & Sons Papers: "The people are still cutting wood and carrying [it] off of the Oaks plantation"; Sprunt to Bellamy, 23 October 1906: "Will you kindly write me a line naming the persons who are trespassing on my property at the Oaks as I would be unable to proceed against them."

105. *Southern Lumberman*, 20 July 1907.

106. See James E. Fickle, *The New South and the "New Competition": Trade Association*

Development in the Southern Pine Industry; Southern Pine Association, *Lumber Awakes!*; Benjamin F. Ulmer, *Lumbermen's Yellow Pine and Cypress Reference Book.*

107. Pinchot and Ashe, "Timber Trees and Forests of North Carolina," 39–41, 119–121, 122–123, 125–126, 131–133, 158, 161, 164, 166, 172, 174, 176, 178.

108. Hugo de Vries, *Naar Californië*, 1–35, especially 3, 4, 6, 8–9, 10–12, 15, 20, 28, 29–30. Translations from Dutch to English are mine.

109. Between 1907 and 1920, the extent of North Carolina's cutover lands decreased from more than 12 million to about 10 million acres, while the portion of the cutover that was not restocking decreased from 19.3 percent to 13.3 percent. Still, by 1907 an average of only 35 percent of the output of sawmills operating in the southeastern corner consisted of longleaf pine lumber; the area's timberlands supported forty-three sawmills with an average daily capacity of more than 30,000 board feet, more than 63 percent of which consisted of second-growth North Carolina or shortleaf pine. See *Directory of Pine Lumber Manufacturers in Virginia, North Carolina, South Carolina, Georgia and Florida.* Shortleaf pine was not a widely dispersed tree in the southeastern corner. According to Pinchot, *Timber Trees*, the area fell outside the shortleaf pine's original growth area. The shortleaf is known, however, to replace longleaf pine once the latter has been cut and cutover lands are routinely burned.

110. *Whiteville News Reporter*, 1925.

111. *Lumberton Argus*, 29 July 1904.

Chapter 4. "After the Lumbermen, What?"

1. Don Marquis, "The Revenges of the Earth," reprinted in the *Southport Herald*, 12 March 1908.

2. O. H. L. Wernicke, *The Lumbermen's Obligation to the South*, 1, 13.

3. *Fayetteville Observer*, 21 February 1828; *Southport Leader*, 4 October 1894.

4. *Southport Standard*, 5 August 1897. The scripture is from Proverbs 3:10 (King James Version).

5. Walter Gilman Curtis, *Reminiscences of Wilmington and Smithville-Southport, 1848–1900* (1905), 61.

6. Allen Maull, *Eastern North Carolina for the Farmer*, 11.

7. Wernicke, *The Lumbermen's Obligation to the South*, 4–6.

8. On republicanism, see Robert W. Tucker, *Empire of Liberty: The Statecraft of Thomas Jefferson*; Gordon S. Wood, *The Radicalism of the American Revolution*; Lacy K. Ford Jr., *Origins of Southern Radicalism: The South Carolina Upcountry, 1800–1860.*

9. Richard White, *Land Use, Environment, and Social Change: The Shaping of Island County, Washington*, 113; Vernon Rosco Carstensen, *Farms or Forests: Evolution of a State Land Policy for Northern Wisconsin, 1850–1932.*

10. Paul M. Gaston, *The New South Creed: A Study in Southern Mythmaking*; Ayers, *The Promise of the New South*, 20–21.

11. Gavin Wright, *Old South, New South*, 7–10, 27–29.

12. Frederick B. Goddard, *Where to Emigrate and Why: Homes and Fortunes in the Boundless West and the Sunny South*, 376–388.

13. North Carolina Land Company, *A Guide to Capitalists and Emigrants*, 11–12, 20, 48

14. North Carolina Literary Board, *Swamp Lands of the State of North Carolina: Facts for Capitalists and Emigrants*. In 1896 the North Carolina Department of Agriculture joined the campaign when it published *North Carolina and Its Resources* to inform outsiders of the immense potential of the state's natural environment.

15. Wright, *Old South, New South*, 38.

16. Advocates of what became known as the "cotton mill movement" or "cotton mill campaign" sought to employ native white laborers in the new mills while keeping blacks on the land as agricultural workers, especially in cotton. See Patrick Hearden, *Independence and Empire: The New South's Cotton Mill Campaign*; Jacqueline Dowd Hall et al., *Like a Family: The Making of a Southern Cotton Mill World*; and Randall Martin Miller, *The Cotton Mill Movement in Antebellum Alabama*.

17. Southern Immigration Association of America, *Proceedings of the Annual Session of the Southern Immigration Association of America*.

18. *Southport Leader*, 20 November 1890.

19. Rowland Berthoff, "Southern Attitudes toward Immigration, 1865–1914," *Journal of Southern History* 17, 328–360.

20. Rupert B. Vance, *Human Geography of the South*, 279.

21. For the rash of communal utopias popping up in the late nineteenth century, see W. Fitzhugh Brundage, *A Socialist Utopia in the New South: The Ruskin Colonies in Tennessee and Georgia, 1894–1901*. For the same frenzy during the 1930s, see Paul K. Conkin, *Tomorrow a New World: The New Deal Community Program*.

22. *Southport Leader*, 28 January 1892.

23. *Southport Leader*, 29 March 1894.

24. Joseph Brown to Minnie McIver, 5 January 1897, folder 5, Brown Family Papers.

25. Glenn F. Strole, *Chadbourn and her Sunny South Colony: A Narrative History*, rev. ed., 10–16; Columbus County, Record of Incorporations, 1887–1938; Ward Little, *Columbus County, North Carolina*, 207–211.

26. Strole, *Chadbourn and her Sunny South Colony*, 35–53; Columbus County, Record of Incorporations, 1887–1938.

27. Tweltfth U.S. Census, 1900, Agriculture; Fifteenth U.S. Census, 1930, Agriculture: strawberry acreage went from 479 acres in 1900 to 2,468 acres in 1930; also *Whiteville News Reporter*, 2 May 1929.

28. *Southport Herald*, 19 April 1905.

29. *Lumberton Argus*, 28 May 1903, 4 June 1903.

30. Twelfth U.S. Census, 1900, Agriculture; Thirteenth U.S. Census, 1910, Agriculture; Fifteenth U.S. Census, 1930, Agriculture: strawberry acreage in Bladen, Brunswick, and Robeson went from forty-one acres in 1900 to 220 acres in 1910, during which year they produced 416,321 quarts for the market. After 1910, strawberry production declined

in the counties outside Columbus, settling at eighty-six acres that yielded 85,838 quarts in 1930.

31. Joseph Brown to Minnie McIver, 29 April 1898, folder 8, Brown Family Papers.

32. Joseph Brown to Minnie McIver, 11 May 1898, folder 9, Brown Family Papers.

33. Joseph Brown to Minnie McIver, 23 December 1897, folder 5; Joseph Brown to Minnie McIver, 3 March 1898, 19 January 1898, folder 7; Joseph Brown to Minnie McIver, 22 April 1898, E. Borden to Ella D. Smith, 26 April 1898, Ella D. Smith to Joseph Brown, 27 April 1898 and 28 April 1898, folder 8, Brown Family Papers.

34. Joseph Brown to Minnie McIver, 26 January 1898; enclosure with letter from T.A. McNeill (of McNeill & McLean) to Joseph Brown, 1 January 1898, folder 7, Brown Family Papers.

35. George F. Lucas to Joseph Brown, 27 June 1898, folder 11, Brown Family Papers (my italics).

36. In 1933 the letterhead of the Southeastern Council read "Purpose: The Logical Development of Mutual Interests" and "Objective: To Advance and Maintain a High State of Civilization in the South." L. H. Duncan to Austin Cary, 14 December 1933, box 7, folder 2, Hugh MacRae Papers.

37. H. A. Cressey to Hugh MacRae, 26 July 1904, box 1378.3, Eric Norden Collection.

38. Hugh MacRae, "To Stop the Back Drift of Farming; to Create Satisfying Rural Life Conditions; to Increase the Purchasing Power of the Farmer—An Objective," in *Proceedings of the Southern Reclamation Conference Held in Washington, D.C., December 14 and 15, 1927*, 70th Congress, 1st Session, Document 45, 39–45.

39. Quoted in Associated Committees on Southern Rural Development, *Southern Rural Life, the Problem, the Solution: Opening the Way for Men to Become Farm Owners*, 22.

40. The literature is extensive, but a good introduction can be found in David L. Carlton and Peter A. Coclanis, eds., *Confronting Southern Poverty in the Great Depression: The Report on Economic Conditions of the South with Related Documents*.

41. On Progressivism in the South, see William A. Link, *The Paradox of Southern Progressivism, 1880–1930*, and Dewey Grantham, *Southern Progressivism: The Reconciliation of Progress and Tradition*.

42. Quoted in *Southern Rural Life*, 22.

43. Hugh MacRae, "Bringing Immigrants to the South: Address Delivered before the North Carolina Society of New York, December 7, 1908," manuscript, Hugh MacRae Papers.

44. Carolina Trucking Development Company (CTDC), *Wealth in Carolina Truck Farms*, pamphlet, Hugh MacRae Papers.

45. Quoted in *Southern Rural Life*, 10.

46. Carl Van Leuven to S. W. Pierce, 20 April 1915, Pierce & Company Papers.

47. CTDC, *Wealth in Carolina Truck Farms*, 9, 22.

48. P. W. Davis, "Soil Map of the Jno. MacRae Estate, 400 A." and "Soil Report upon

the Jno. MacRae Estate, 400 A.," 3 April 1906, Hugh MacRae Papers. The engineer conducting the survey considered the tract of 400 acres belonging to the John MacRae estate south of Leland in Brunswick County, despite being largely unimproved and in need of drainage, "very well located"; moreover, he noted that the Southport fine sand that made up the largest portion of the soil was suitable for growing celery. See also soil map of "John MacRae 1532 A. Tract" and "Soil Report of the John MacRae Tract," 8 September 1905; soil map of "F. W. Weeks 400 A. Tract" and "Soil Report of the F. W. Weeks Tract," 18 August 1905; "Lands in Pender and Brunswick Counties," undated typed manuscript; "Official Map of Artesia Colony, Columbus County, 1905–1908," with marginalia dated 1905 and 1920; and "Brief Description of Properties Owned or Controlled by Hugh MacRae & Company," typed manuscript, 1937, Hugh MacRae Papers.

49. "Feeding Hogs in the South," typed manuscript, Hugh MacRae Papers.

50. J. A. Taylor to James Sprunt, 29 September 1906, box 2, folder 2, Alexander Sprunt & Sons Papers (my italics).

51. Hugh MacRae, "Bringing Immigrants to the South"; David S. Cecelski and Timothy B. Tyson, ed., *Democracy Betrayed: The Wilmington Race Riot of 1898 and Its Legacy*, 5; H. Leon Prather, *We Have Taken a City: Wilmington Racial Massacre and Coup of 1898*, 49–80.

52. George Byrne, "Hugh MacRae's Practical Application of Common Sense in Colonization," *Manufacturers Record*, 30 May 1912.

53. Robert W. Vincent, "Successful Immigrants in the South," *The World's Work*, November 1908, 10908–10911.

54. Hugh MacRae, "Colonization Opportunity in the Southern States," undated typed manuscript, Hugh MacRae Papers. There is some disagreement among observers whether Artesia was for Englishmen or Dutchmen. According to *The World's Work*, the English immigrants had failed to live up to expectations, after which MacRae began to court the Dutch.

55. Quoted in John Faris Corey, "The Colonization and Contributions of Emmigrants Brought to Southeastern North Carolina by Hugh MacRae," 22; Hugh MacRae, "Human Engineering," in Southern Land Congress, *Soldier Settlements in the South* (1918), 97–101.

56. Donald Denoon and Philippa Mein-Smith with Marivic Wyndham, *A History of Australia, New Zealand and the Pacific*, 227–228. Eric Pawson and Tom Brooking, eds., *Environmental Histories of New Zealand*, especially Geoff Park, "'Swamps which might doubtless Easily be drained': Swamp drainage and its impact on the indigenous," 151–165.

57. Lower Cape Fear Historical Society Inc., "The Need for Greener Pastures," "Hugh MacRae's Dream," "The First Try," and "The Jewish Farmers," *Bulletin* 40, 4–11; Marianne Mooijweer, *De Amerikaanse Droom van Frederik van Eeden*; Carolina Trucking Development Company, *De Van Eeden Kolonie*, (pamphlet 1913), Hugh MacRae Papers.

58. See James R. Scobie, *Revolution on the Pampas: A Social History of Argentine Wheat, 1860–1910*; Thomas E. Skidmore, *Black into White: Race and Nationality in Bra-*

zilian Thought; and Thomas H. Holloway, *Immigrants on the Land: Coffee and Society in São Paulo, 1886–1934*.

59. G. E. di Palma-Castiglione to Hugh MacRae, 21 April 1910, box 6, folder 3, Hugh MacRae Papers.

60. See Robert Gough, *Farming the Cutover: A Social History of Northern Wisconsin, 1900–1940*; Alan Helgeson, *Farms in the Cutover: Agricultural Settlement in Northern Wisconsin*; Carstensen, *Farms or Forests*; and James Kates, *Planning a Wilderness: Regenerating the Great Lakes Cutover Region*.

61. White, *Land Use, Environment and Social Change*, 113–141.

62. Liberty H. Bailey, *The Country-Life Movement in the United States* (1911), is a contemporary description of the movement; William L. Bowers, *The Country Life Movement in America, 1900–1920*, gives a detailed history; Mooijweer, *De Amerikaanse Droom van Frederik van Eeden*, puts MacRae in the context of the movement's history. The distinction between "urban agrarians" and "anti-agrarians" is from David B. Danbom, *The Resisted Revolution*, 29. All suggest a general ambivalence among reformers about the proper place of city and country in modern society. MacRae's company qualified its praise for the city's many facilities with disdain for the lives of people who lived and worked there, comparing their unenviable conditions with the "absolute independence" truck farms offered their prospective settlers: "Contrast this condition of independence with that of the city man—the mechanic, the factory or mill worker. There are no strikes—no loss of employment from hard times, or the displeasure of the boss. The farmer goes about his work . . . without that feeling of drudgery which 'the whistle' invariably produces. Instead of the whistle to call him to work, it requires the 'dinner bell' to bring him from the field." CTDC, *Wealth in Carolina Truck Farms*. MacRae, in "Bringing Immigrants to the South," also argued that drawing immigrants from rural Europe to American farms would stave off an influx of undesirable urban immigrants as agriculture "separates the scum of the European cities from those who have led the rugged, wholesome life of farmers."

63. See David McCally, *The Everglades: An Environmental History*, 106–128, and Ann Vileisis, *Discovering the Unknown Landscape*, 111–141.

64. H. H. Brown to W. G. Lewis, 29 November and 14 December 1885, box 6, folder 10, North Carolina State Board of Education, wrote that he was working with a camp on the swamp in Bug Hill and referred to an earlier letter he sent from Jones Lake; J. F. Ranco wrote from Camp Buies, where he was at work in the swamp, 8 June 1886, box 6, folder 12. See also "Report on Big Colly Bay, Bladen County, Area 2166 Acres," Bladen County File No. 15, undated typed manuscript, box 1378.6, Eric Norden Collection. This report includes letters from Eric Norden to E. C. Brooks, 4 October 1920, and from Brooks to Norden, 18 October 1920. W. C. Kerr, "Report on the Swamp Lands," undated typed manuscript, but marked "Read at Literary Board meeting May 25, 1866, box 13, Board of Education.

65. *Congressional Record: House*, 25 March 1908, 3905–3907; also *Southport Herald*, 2 April 1908.

66. *Southern Lumberman* 27, 17 August 1907, 7 September 1907, and 26 October 1907.

67. *Southern Lumberman* 36, 9 September 1916, 2 December 1916, and especially 16 December 1916, an issue largely devoted to cutover land. Among the stories it ran were those of the Tennessee Valley Iron and Railroad Company, which used the journal's pages to promote the raising of Angora goats on "well-drained cutover lands" as "the COMING BIG THING" and the sale of its lands "on easy terms to thrifty farmers . . . from the North and Northwest."

68. Southern Pine Association, *Southern Pine and Reconstruction*.

69. Wernicke, *Lumbermen's Obligation*, 7, 14–19, 21–24, 27–31. See also Southern Land Congress, *Soldier Settlements in the South*, for a discussion of plans to settle soldiers returning from World War I on cutover land.

70. Wernicke, *Lumbermen's Obligation*, 8, 31, 33–45. Unlike other reformers, Wernicke did not foresee a significant role for immigrants in this landscape transformation. Although he admitted that the colonization of immigrants in farm communities, which had gained adherents among his fellow entrepreneurs, presented "alluring prospects," he doubted that immigration to the southern states would increase to such an extent that the bulk of the cutover could be turned into farms. The agricultural undertaking he envisioned for the cutover—ranching that required extensive land use but limited labor—eliminated the need to engage in the active recruitment of farmers from outside the region: "Covering the land with flocks and herds" would "of itself attract widespread attention among thoughtful and desirable people." Wernicke nonetheless hinted at a social initiative typical of the era: "Progressive encouragements" would turn southern blacks, whose "greater birth rate" and "fitness to perform manual labor under the prevailing climatic and social conditions" already made them desirable workers, into the primary source of labor for the sheep farming venture. Indeed, he proclaimed, "the raising of more and better negroes in the South is as much to be desired as the raising of good animals."

71. Wade H. Phillips to Howard Elliott, 22 December 1926, box 6, ABCD 1926–1935.

72. C. B. Williams to Howard Elliott, 10 December 1926, box 6, ABCD 1926–1935; C. B. Williams, "A Safe Cropping Program for North Carolina Farmers to Follow in Growing Cotton in 1927," typed manuscript, box 6, ABCD 1926–1935; *Whiteville News Reporter*, 19 August 1926.

73. See Gilbert Fite, *Cotton Fields No More: Southern Agriculture, 1865–1980*, 80–81, on Seaman Knapp; also and Joseph Cannon Bailey, *Seaman Knapp: Schoolmaster of American Agriculture*.

74. Andrew Needham, "'Qualified to Succeed': Elwood Mead and the Durham Agricultural Colony, 1915–1919," *Ex Post Facto* 5 (1996).

75. Elwood Mead, "Plan for the Creation of Organized Rural Communities in the South," undated typed manuscript, box 6, ABCD 1926–1935.

76. "Report on Southern Reclamation Conference, Washington, D.C., January 24,

1928," and Wade H. Phillips to A. W. McLean, 25 January 1928, box 6, ABCD 1926–1935. Conferences were held on 14–15 December 1927, and on 21 January 1928. See Department of the Interior, *Reclamation and Rural Development in the South*, House of Representatives, 69th Congress, 2nd session, document no. 765, part 1, 24–28.

77. For example, Finkenstaedt was at the Southern Land Congress to applaud heartily the prospect of settling soldiers on the cutover: F. L. Finkenstaedt, "The Carolinas," in *Soldier Settlements in the South*, 88–90.

78. "Colonization Questionnaire" and letter, John M. Merriam to Elwood Mead, 28 August 1926, box 6, ABCD 1926–1935.

79. W. B. Cobb, "Questionnaire on Pate Estate," 15 November 1926, box 6, ABCD 1926–1935; Wade H. Phillips to C. B. Williams, 26 October 1926, and C. B. Williams to Wade H. Phillips, 3 November 1926, box 6, ABCD 1926–1935. *Reclamation and Rural Development in the South*, 21–22.

80. "Report on Southern Reclamation Conference, Washington, D.C., January 24, 1928;" Wade H. Phillips to A. W. McLean, 25 January 1928, box 6; Elwood Mead, "Plan for the Creation of Organized Rural Communities in the South," undated typed manuscript, box 6, ABCD 1926–1935. Mead, "Beginning of Planned Rural Development in the United States," in *Soldier Settlements in the South*, 23–33

81. Printed in *Southport Herald*, 12 March 1908.

82. Vincent, "Successful Immigrants in the South," *The World's Work*, November 1908, 10908–10911.

83. Byrne, "Hugh MacRae's Practical Application of Common Sense in Colonization," *Manufacturers Record*, 30 May 1912.

84. CTDC, "How to Own a Truck Farm in the Eastern North Carolina Colonies," Hugh MacRae Papers; Vincent, "Successful Immigrants in the South."

85. Columbus County, Record of Incorporations, 1897–1938. The five stockholders were A. Van Vuuren, M. J. Bast, J. J. Van Wouw, J. H. Van Elk, and H. Brinksma—each held one share worth $20.

86. Fourteenth Census of Population, 1920, North Carolina, vol. 22, Columbus County; "Official Map of Artesia Colony, Columbus County" and "Acres Sold during Year 1920—Artesia Colony," typed manuscript, Hugh MacRae Papers; Ward Little, *Columbus County, North Carolina*, 91.

87. Fourteenth Census of Population, 1920, North Carolina, vol. 22, Columbus County. Delco was home to a single man from Hungary; a family of three from New York; a single man from Holland; a German family of five whose three kids were all born in North Carolina; a family from Maryland; a North Carolina family of three whose daughter was born in Poland; a man from Michigan whose parents were born in England and whose wife and son were born in Ohio; an Austrian man and his German wife whose three children were born in Pennsylvania (1) and North Carolina (2); a German man and his North Carolina-born wife and two sons; and a Hungarian couple whose son was born in Canada and whose daughter was born in Ohio. All immigrants were farmers. See also "Brief Description of Properties Owned or Controlled by Hugh Mac-

Rae & Company," typed manuscript, 1937, Hugh MacRae Papers; various maps dated 1905–1908, Eric Norden Collection; Ward Little, *Columbus County, North Carolina*, 349.

88. Eleventh U.S. Census, 1890, Agriculture; Twelfth U.S. Census, 1900, Agriculture; Thirteenth U.S. Census, 1910, Agriculture; Fourteenth U.S. Census, 1920, Agriculture; Fifteenth U.S. Census, 1930, Agriculture. Between 1900 and 1930 the combined acreage for vegetables increased from 2,419 acres to 2,791 acres (a 15.4% increase), while the yield increased from $139,070 to $174,385 (25.4%) or from $57 to $62 per acre (8.8%). Farmers also expanded acreage for white potatoes, an alternative to the much more common sweet potatoes, from 90 acres in 1890 to 1,307 acres in 1930 (1,352.2%). Yield increased by 1878.9 percent from 5,251 bushels (58.3 bushels per acre) to 103,912 bushels (79.5 bushels per acre—a 36.4 percent increase in productivity).

89. Tenth U.S. Census, 1880, Agriculture; Eleventh U.S. Census, 1890, Agriculture; Twelfth U.S. Census, 1900, Agriculture; Thirteenth U.S. Census, 1910, Agriculture; Fourteenth U.S. Census, 1920, Agriculture; Fifteenth U.S. Census, 1930, Agriculture (sheep not listed in 1930). Corn yields, a staple food for humans but also crucial for raising hogs, increased from 1,001,861 bushels on 113,566 acres in 1890 to 2,268,970 bushels on 130,124 acres in 1930. Between 1880 and 1920, acreage for hay and forage crops went from 417 to 40,304 acres, with the biggest increases occurring between 1880 and 1890 (469%) and between 1910 and 1920 (309%); yield increased from 324 tons to 23,652 tons (7200%). Hay crop decline 1920–1930: from 23,162 to 21,864 acres.

90. Fourteenth U.S. Census, 1920, Agriculture; Fifteenth U.S. Census, 1930, Agriculture. 68.8 percent of Columbus County farmers owned their farms—compared to 80.8 percent in 1920. In Bladen County, farm ownership equaled 72.4 percent, down from 73.3 percent a decade earlier. In Brunswick, 81 percent of farmers—down from 1920s 85.4 percent—still operated farms they could call their own.

91. Elwood Mead to Wade H. Phillips, 16 December 1926, and "Tentative Itinerary of Commission on Reclamation and Rural Development," typed manuscript, box 6, ABCD 1926–1935.

92. A. F. Corbin, "Preliminary Report, Proposed Supervised Settlement 100 Families: Pembroke, Robeson Company, North Carolina," typed manuscript, and Corbin to O. O. Dukes, 4 December 1926, box 6, ABCD 1926–1935.

93. Elwood Mead to Wade H. Phillips, 9 June 1927, box 6, ABCD 1926–1935.

94. Phillips to Mead, 14 June 1927, Cardwell to H. Phillips, 6 July 1927, Phillips to Williams, 6 July 1927, box 6, ABCD 1926–1935.

95. *Proceedings of the Southern Reclamation Conference Held in Washington, D.C., December 14 and 15, 1927*, iv.

96. Allen Maull, *Eastern North Carolina for the Farmer*, 5, 6, 8, 11, 21, 28.

97. "Abstract of Title of Bladen County Lumber Company," box 1378.9, Eric Norden Collection.

98. Fifteenth U.S. Census, 1930, Agriculture. In 1930, of 16,168 farms that reported cutting forest products for home use and sale, 9,336 (58%) produced 97,105 cords of fire-

wood—the product most likely to be used at home. Only 695 farms (4%) engaged in the production of saw and veneer logs (27,894,000 board feet). A mere 158 (1%) produced railroad ties, 98 (0.6%) engaged in the production of fence posts, 27 (0.2%) made poles and piling, and only one reported producing pulpwood. Conversely, a growing number of farmers resorted to buying lumber and turpentine for domestic use, and an even larger increase occurred among farmers and their offspring who took up employment as seasonal boxers, loggers, and sawyers to alleviate financial stress.

99. Tenth U.S. Census, 1880, Agriculture; Thirteenth U.S. Census, 1910, Agriculture; Fifteenth U.S. Census, 1930, Agriculture. Wooded farmland comprised 1,342,066 acres in 1880, 924,897 acres in 1910, and 475,455 acres in 1930.

100. L. H. Duncan to Austin Cary, 14 December 1933, box 7, folder 2, Hugh MacRae Papers.

101. Nelson MacRae to P. J. A. Berckmans, 9 February 1931 (two separate letters); Nelson MacRae to James E. Whitin, 9 February 1931; James E. Whitin to Nelson MacRae; John D. Bellamy to Nelson MacRae, 17 February 1931; Nelson MacRae to P. J. A. Berckmans, 19 February 1931; Nelson MacRae to F. L. Finkenstadt, 19 February 1931; Nelson MacRae to John Bellamy, 14 April 1931; J. W. Jackson to Nelson MacRae, 25 June 1931; Nelson MacRae to Wayne Cunningham, 7 September 1931, box 7, folder 1, Hugh MacRae Papers. Wayne Cunningham to Dick Reynolds, 24 October 1932; Nelson MacRae to Wayne Cunningham, 11 December 1933, box 7, folder 2, Hugh MacRae Papers.

102. K. B. Council and Brown's wife Minnie were also engaged in a history or genealogy project regarding families of the area, further cementing ties between community leaders and their home region. K. B. Council to Mrs. J. A. Brown, 18 November 1912, 2 December 1912, 25 December 1912, folder 24; K. B. Council to Mrs. Brown, 14 March 1913, folder 25; K. B. Council to Mrs. Brown, 24 July 1917, folder 29, Brown Family Papers.

103. Joseph Brown to Minnie McIver, 5 June 1898, folder 11, Brown Family Papers.

104. Joseph Brown to Minnie McIver, 23 December 1897, folder 5, Brown Family Papers.

105. Joseph Brown to Minnie McIver, 5 November 1897; Joseph Brown to Minnie McIver, 28 February 1898, folder 6, Brown Family Papers.

106. Joseph Brown to Minnie McIver, 17 January 1898, folder 6, Brown Family Papers.

107. Alvin Johnson to Arthur Flatow, 3 January 1940; H. M. Pinckney to Arthur Flatow, 14 June 1940; Arthur Flatow to H. M. Pinckney, 12 July 1940; Alvin Johnson to Arthur Flatow, 18 September 1940; Alvin Johnson to Arthur Flatow, 11 October 1940, box 1, file 3, Van Eden Papers.

108. See W. G. Wahlenberg, Greene, and Reed, *Effects of Fire and Cattle Grazing on Longleaf Pine Lands, as Studied at McNeill, Miss.*; also Elinor Melville, *A Plague of Sheep*.

109. Wernicke, *Lumbermen's Obligation*, 14.

110. *Southern Rural Life*, 11.

Chapter 5. "It Would Seem Necessary to Restrict Such People"

1. F. L. Finkenstaedt to Eric Norden, 3 October 1934; Norden to Finkenstaedt, 31 August 1935 and 19 May 1936; Norden to U.S. Senator Josiah W. Bailey, 17 July 1934 and 21 January 1935; Norden to Carl C. Taylor, 14 November 1934; Taylor to Norden, 28 September 1934; Norden to Joseph Kircher, 11 October 1934; Kircher to Nelson MacRae, 21 November 1934, box 1378.32, Eric Norden Collection.

2. Samuel Hays, *Conservation and the Gospel of Efficiency: The Progressive Conservation Movement, 1890–1920.*

3. John F. Reiger, *American Sportsmen and the Origins of Conservation.*

4. Foresters at George Vanderbilt's Biltmore estate in Asheville, North Carolina, were among the first to implement the scientific forestry techniques that Gifford Pinchot would turn into federal policy as Roosevelt's head forester. In 1890 the state created the Geological and Economic Survey, later renamed the Department of Conservation and Development, which included the Geological Survey, the Forestry Division, and the Division of Game and Fisheries, all agencies that dedicated themselves to surveys of natural resources, proposals on methods of necessary protection and profitable exploitation, and the enforcement of commensurate laws. For a rather celebratory account of this legacy, see O. C. Goodwin, *Eight Decades of Forestry Firsts.*

5. Margaret Lynn Brown, *The Wild East: A Biography of the Great Smoky Mountains*; Richard White, *The Organic Machine*; Joseph E. Taylor III, *Making Salmon*; Karl Jacoby, *Crimes against Nature.*

6. Hays, *Conservation and the Gospel of Efficiency,*and *Explorations in Environmental History: Essays*; Max Oelschlaeger, *The Idea of Wilderness: From Prehistory to the Age of Ecology*, 209–212.

7. Condemning the creation of a "new, stripped-down forest" that consisted largely of "stands of same-age trees in linear alleys," James C. Scott, *Seeing like a State*, 11–52, accuses states of sacrificing forest diversity and resilience for the sake of a manageable uniformity that undermines the forest's health.

8. *Southport Standard*, 8 March 1900.

9. *Whiteville News Reporter*, 5 March 1925.

10. *The Lumbermen's Credit Rating Book: Clancy's Red Book and Lumbermen's Blue Book of the Lumbermen's Red and Blue Book Service*, 112th ed.

11. *Whiteville News Reporter*, 6 March 1924, 24 April 1924.

12. *Whiteville News Reporter*, 29 January 1925, 19 February 1925, 21 May 1925, and 18 June 1925.

13. *Whiteville News Reporter*, 16 February 1928.

14. North Carolina Department of Conservation and Development Forest Service, "Forest Fire Control in North Carolina," typed manuscript, 1938, box 13, Activities of the Department of Conservation and Development (ADCD) 1936–1941.

15. Thomas Clark, *The Greening of the South*, 104–108; Jack Porter Oden, "Development of the Southern Pulp and Paper Industry, 1900–1970," 107–142.

16. Various land deeds, dated between 1927 and 1931, box 1378.6; Charles Luke to Eric Norden, 21 May 1924, box 1378.31; deed, Waccamaw Lumber Company to Cooper River Timber Company, 26 May 1927, in "Abstract of Title: Waccamaw Lumber Company," box 1378.20; Charles Luke to Eric Norden, 3 September 1925, Norden to Luke, 26 October 1925 and 21 May 1926, and Norden to S. C. Sweeny, 17 January 1927, box 1378.31, Eric Norden Collection.

17. C. F. Evans to C. O. Brown, 10 October 1941, box 13, ABCD 1936–1941; Columbus County Index to Real Estate Conveyances: Supplement Grantees 1923–1966, vol. C; Bladen County Index to Real Estate Conveyances: Grantees 1784–1961, vol. C; Robeson County Index to Real Estate Conveyances: Grantee 1787–1939, vol. C.

18. I. F. Eldredge, "The Naval Stores Industry: How It Can Be Perpetuated in the South"; Mark V. Wetherington, *The New South Comes to Wiregrass Georgia, 1860–1910*; Robert Outland, "Another New South: Patterns of Continuity in the Southern Naval Stores Industry."

19. *Lumberton Argus*, 15 January 1903.

20. *Southport Leader*, 17 April 1890, 5 June 1890, 27 November 1890, and 22 March 1894; *Southport Standard*, 8 March 1900; *Southport Herald*, 20 February 1908, and 11 June 1908. K. Clyde Council of the Council Tool Company wrote State Forester J. S. Holmes that George P. Shingler of the *Naval Stores Review* visited several turpentine plants near Wilmington in August of 1930. Council to Holmes, 4 September 1930, box 8, ABCD 1925–1936.

21. *Whiteville News Reporter*, 27 March 1924, 31 March 1927, 8 December 1927, and 16 February 1928.

22. Goodwin, *Eight Decades of Forestry Firsts*, 7.

23. *Southport Leader*, 20 March 1890, 27 November 1890, 22 October 1891, 4 February 1892, 22 December 1892, 26 December 1895, 27 August 1896, and 22 October 1896; *Lumberton Argus*, 17 December 1903; *Southport Standard*, 3 November 1904.

24. *Whiteville News Reporter*, 19 November 1925, 14 October 1926, 18 November 1926, 6 January 1927, 1 September 1927, 8 December 1927, 13 September 1928, and 20 December 1928.

25. *Whiteville News Reporter*, 29 July 1926, 7 July 1927, 18 April 1929, and 29 May 1930; *Southport Leader*, 12 June 1890; *Lumberton Argus*, 13 December 1904; *Southport Herald*, 4 June 1908 and 11 June 1908.

26. North Carolina Department of Conservation and Development, "Timber Survey," handwritten manuscript, box 4; untitled manuscript, box 3; W. A. Peterson, timber descriptions of Bladen, Brunswick, Columbus, and Robeson counties, handwritten and typed manuscripts, box 5, ABCD 1926–1935.

27. North Carolina Department of Conservation and Development, "Timber Survey," handwritten manuscript, and "Timber Resources of North Carolina: Total Estimated Stand by Districts in Percentage and Thousand Board Feet," box 3, ABCD 1926–1935; *Whiteville News Reporter*, 27 February 1930.

28. North Carolina Department of Conservation and Development, "Report of Game Killed in North Carolina during Season 1927–28." The actual numbers of animals killed were likely much higher than those reported as some 3,000 holders of hunting licenses in the four counties did not report their harvest.

29. "Periodic Catch of Fish in Waccamaw Lake in Percentages, 1930–31," typed manuscript, box 8, ABCD 1926–1935.

30. *Whiteville News Reporter*, 6 August 1925, 27 August 1925, 10 June 1926, 26 July 1928, 27 September 1928, 11 October 1928, 18 July 1929, 8 August 1929, and 7 August 1930.

31. *Southport Leader*, 23 November 1893, 11 October 1894, and 2 July 1896; *Whiteville News Reporter*, 7 August 1924, 17 September 1925, 10 December 1925, 9 September 1926, 18 August 1927, 25 September 1930, and 27 November 1930.

32. *Southport Leader*, 26 January 1891 and 22 August 1895; *Whiteville News Reporter*, 9 June 1927, 11 August 1927, 25 July 1929, 1 August 1929, and 13 August 1931.

33. *Southport Leader*, 25 August 1892, 10 November 1892, and 15 August 1895.

34. *Whiteville News Reporter*, 17 October 1929 and 9 October 1930; *Southport Standard*, 15 March 1900; *Lumberton Argus*, 12 May 1905.

35. *Whiteville News Reporter*, 27 March 1930; another "huge" wildcat was reported near Delco on 19 June 1930.

36. *Whiteville News Reporter*, 23 August 1928.

37. *Whiteville News Reporter*, 27 August 1931.

38. *Lumberton Argus*, 18 October 1904; *Whiteville News Reporter*, 7 March 1929.

39. *Whiteville News Reporter*, 7 August 1924.

40. *Whiteville News Reporter*, 19 July 1928, 31 January 1929, and 28 February 1929.

41. *Whiteville News Reporter*, 11 August 1927. The *News Reporter* routinely reported on the discovery of stills and the capture of their operators. The individual instances are too numerous to mention, but the newspaper made explicit mention of the location of stills on 3 January 1929, 28 February 1929, and 20 June 1929.

42. *Lumberton Argus*, 28 January 1904, 11 February 1904, and 17 March 1904.

43. Hays, *Conservation and the Gospel of Efficiency*, 27–48, quote on 40.

44. Sherry Olson, *The Depletion Myth*.

45. Howard Odum, *Southern Regions of the United States*, 21, 39–55, used the term "waste" to describe many of the South's socioeconomic deficiencies, but saw waste of physical resources as central to this multitude of problems.

46. J. S. Holmes, "Autobiographical," typed manuscript, 1931, box 13, ABCD 1926–1935.

47. *Whiteville News Reporter*, 26 June 1924.

48. W. A. Peterson, timber descriptions, counties, handwritten and typed manuscripts, box 5, ABCD 1926–1935.

49. See Michael Thompson, "High on the Hog: Swine as Culture and Commodity in Eastern North Carolina"; Kirby, *Poquosin*, 98–105, 205–207.

50. *Whiteville News Reporter*, 5 January 1928; John Reiger, "Wildlife, Conservation, and the First Forest Reserve," in Char Miller, ed., *American Forests: Nature, Culture, and Politics*, 35–47.

51. Unknown author to Walter Pinckney High, 26 October 1946, box 4, File 27, Oscar High Papers.

52. *Whiteville News Reporter*, 6 October 1927.

53. J. S. Holmes, "Memorandum to the Director," 3 February 1928, box 3, ABCD 1926–1935. John Koonce to Oscar High, 2 July 1923; Walter G. Sheppard to Oscar High, 19 June 1923, box 4, file 2; D. Jerrigan to Oscar High, 4 June 1923, box 5, Oscar High Papers.

54. *Lumberton Argus*, 19 August 1904.

55. "A Bill to Be Entitled An Act to Prevent the Alienation of State Lakes," undated typed manuscript; J. S. Holmes, "Legal Basis upon Which the North Carolina Department of Conservation and Development Has Taken Over the Administration of the State Lakes: Consolidated Statutes of North Carolina 1920, State Lands, Chapter 128," typed manuscript, 22 February 1926; Dennis G. Brummitt to J. S. Holmes, 12 August 1929, box 8, ABCD 1924–1929; "Regulations Dealing with the Public Use of the State Lakes," undated typed manuscript, box 8, ABCD 1928–1935.

56. For instance the North State Game Club, a hunting club led by Maxton's leading businessman R. L. McLeod, purchased from Eric Norden "the exclusive right to hunt, kill, and carry away any and all game or wild life" on a 4,700-acre tract in Bladen County, as well as "the right and privilege to post and patrol for the purpose of protecting game and wild life." Indenture, 28 October 1926, box 1378.6, Eric Norden Collection. J. D. McAllister of Lumberton served as president of the Cape Fear Gun Club, whose members hunted on 16,000 acres of private land in Bladen, while Dr. R. B. Whitaker of Whiteville led the Hickory Mount Club, which hunted on 10,000 acres of private land in Columbus County. See North Carolina Department of Conservation and Development, *Hunting in North Carolina*, 28–29. The Bellamy or "Grovely" Plantation in Brunswick County was a piece of private property that several interested parties thought should be turned into a game preserve. Once again, it was people like F. L. Finkenstaedt and Nelson MacRae who set about locating a private or government buyer: Finkenstaedt to MacRae, 26 February 1931; MacRae to Finkenstaedt, 27 February 1931; Finkenstaedt to MacRae, 28 February 1931; MacRae to Wayne Cunningham, 13 November 1931; Wayne Cunningham, "Bellamy Plantation," undated advertisement; U.S. Department of Agriculture, "Proposal for Sale of Land," 15 March 1934, Hugh MacRae Papers.

57. *Lumberton Argus*, 11 December 1902.

58. *Whiteville News Reporter*, 9 October 1924 and 21 August 1930.

59. *Whiteville News Reporter*, 22 October 1925.

60. *Whiteville News Reporter*, 5 May 1927.

61. *Southern Lumberman*, 2 October 1907.

62. *Whiteville News Reporter*, 18 June 1925.

63. "Last Year's Forest Fires in Bladen County," Press Notice No. 93, 14 September 1917, box 99, North Carolina Geological and Economic Survey (NCGES), Forestry Division.

64. *Whiteville News Reporter*, 5 June 1930.

65. *Whiteville News Reporter*, 26 January 1928.

66. *Lumberton Argus*, 8 April 1904.

67. J. A. McLean to Joseph Hyde Pratt, 1 December 1922, box 105, North Carolina Department of Conservation and Development, Geological and Economic Survey (NCDCD-GES).

68. *Whiteville News Reporter*, 12 January 1928.

69. Joseph Hyde Pratt to J. L. Underwood, 20 April 1922; Pratt to J. M. Sherwood, 8 November 1922, box 105, NCDCD-GES.

70. *Whiteville News Reporter*, 22 October 1925.

71. *Whiteville News Reporter*, 30 October 1930.

72. See Stephen Pyne, *Fire in America: A Cultural History of Wildland and Rural Fire*.

73. *Southport Leader*, 17 March 1892.

74. *The Diary of Elizabeth Ellis Robeson*, entries for 12 April 1860, 13 April 1860, and 25 July 1861.

75. NCGES, *Economic Paper* No. 25. An exception was the case of Rufus Coleman of Brunswick County, who in 1893 was charged with burning the woods. *Southport Leader*, 14 September 1893.

76. *Whiteville News Reporter*, 8 January 1925.

77. *Whiteville News Reporter*, 5 June 1930.

78. *Whiteville News Reporter*, 21 August 1930.

79. *Whiteville News Reporter*, 5 June 1930. Bladen, Brunswick, and Robeson paid $1,500, $2,000, and $2,520 respectively.

80. *Whiteville News Reporter*, 26 June 1924.

81. *Whiteville News Reporter*, 17 February 1927.

82. *Whiteville News Reporter*, 29 October 1925, 10 March 1927, 5 May 1927, 12 May 1927, 29 September 1927, and 8 December 1927.

83. "Lumbermen's Conference at Goldsboro, N.C.," 11 November 1920, box 105, NCDCD, GES.

84. "Monthly Report of Wm. D. Clark, Chief Forest Fire Warden, North Carolina Geological and Economic Survey, November 1920," box 105; "Monthly Report of the State Forester to the State Geologist, No. 1920," box 99, NCDCD-GES. See also "Organization of Co-Operative Forest-Fire Protective Areas in North Carolina, Being the Proceedings of the Special Conference on Forest Fire Protection Held as Part of the Conference on Forestry and Nature Study, Montreat, N.C., July 8, 1915," in NCGES, *Economic Paper* No. 42.

85. *Whiteville News Reporter*, 3 September 1925 and 18 August 1927.

86. *Whiteville News Reporter*, 4 October 1928.

87. Deed, Cooper River Timber Company to N.C. Department of Conservation and Development, 24 September 1930, box 1378.6, Eric Norden Collection.

88. *Whiteville News Reporter*, 14 January 1926.

89. See http://www.counciltool.com/forest.html.

90. *Whiteville News Reporter*, 18 December 1930.

91. "Bear Pen," manuscript (no date, no author), box 3, folder 270/3/1, "Industry," Lake Waccamaw History.

92. W. B. Keziah to J. S. Holmes, 24 January 1938; J. S. Holmes, "Report to the Board of Conservation and Development, Division of Forestry, Jan. 1–June 30, 1939" and "Semi-Annual Report of State Forester to the Board of Conservation and Development, January 13–14, 1941," box 14, ABCD 1936–1941.

93. George R. Foulke Jr., "Notes of Conversation at E.C.W. Headquarters at Southport, N.C., April 16, 1935," box 6, ABCD 1926–1935; "Activities of the CCC for the period July 1 to December 31, 1937," box 14, ABCD 1936–1941; McCormick to McCloskey, 1 September 1934, box 6, ABCD 1926–1935; J. S. Holmes, "Memorandum to the Director," 14 February 1936, box 14, ABCD 1936–1941.

94. Holmes to Cardwell, 19 April 1933, box 8, ABCD 1928–1935.

95. Civilian Conservation Corps Camp Papers, Company 4482, *Backwoods Echo*, Elizabethtown, N.C.

96. Zane Grey, *The Young Forester*, 24.

97. The Coca-Cola Bottling Company, *Our America. Lumber: Trees, Timber, Wood Products*; poster, U.S. Forest Service, "Smokey says—burned timber builds no homes: prevent forest fires"; *Girl Scout Handbook*, rev. ed., 84–110, 135, 136, 137–8, 146, 162, 209.

98. J. G. Formy-Duval, "Forest Tree Nursery and Genetics Operations and Facts," unpublished manuscript on Federal Paper Board letterhead, 21 January 1972.

99. J. S. Holmes, "Memorandum to the Director," 23 April 1928, box 3, NCDCD.

100. *Petition for Condemnation, United States of America, Petitioner vs. 1,000.00 acres, more or less, of land in Bladen County, North Carolina, Charles Brown, et al., Defendants*," 26 August 1937, box 14, ABCD 1936–1941.

101. A. M. Council to J. S. Holmes, 29 July 1929; J. S. Holmes to A. M. Council, 30 July 1929; W. A. Holder to J. S. Holmes, 3 August 1929 and 8 August 1929; W. C. Ewing to Holmes, 11 September 1929; Holmes to Ewing, 19 September 1929, box 8, ABCD 1924–1929.

102. J. D. Beatty to J. S. Holmes, 10 April 1928; J. S. Holmes to W. A. Ferguson, 14 April 1928; W. A. Ferguson to J. S. Holmes, 10 April 1928; Wade H. Phillips to J. S. Holmes, 3 May 1928; Holmes to Phillips, 5 May 1928; Holmes to Phillips, 14 May 1928, box 8, ABCD 1924–1929; *Bladen Journal*, 19 April 1928.

103. J. S. Holmes, "Report to the Board of Conservation and Development, Division of Forestry, Jan. 1–June 30, 1939"; NCDCD, Forestry Division, "Report to the Board, Jan. 1–July 1, 1940," box 14, ABCD 1936–1941.

104. "Report to the Board for the Period July 1, 1939–January 1, 1940," typed manuscript, box 14, ABCD 1936–1941.

105. "Regulations Governing Administration of the State Lakes," typed manuscript, 1932, box 8, ADCD 1926–1935.

106. NCDCD, Forestry Division, "Report to the Board, Jan. 1–July 1, 1940," box 14, ABCD 1936–1941.

107. J. S. Holmes, "Memorandum to the Director: Land Bordering State Lakes," 23 January 1928; K. Clyde Council to J. S. Holmes, 2 April 1928; J. S. Holmes, "Memorandum to the Director: State Game Refuges, Forests and Parks," 13 June 1928, box 3, ABCD 1928–1935. J. S. Holmes to Cameron MacRae, 23 February 1928; Hugh MacRae to R. Bruce Etheridge, 5 December 1934, box 8, ABCD 1928–1935. MacRae gave a strip of land to the Department of Conservation and Development for the establishment of a state park at Fort Fisher, Brunswick County.

108. J. S. Holmes, "Memorandum to the Director: Jones and Singletary Lakes," 29 August 1938; Holmes, "Memorandum to Committee on Forestry and Parks, N.C. Board of Conservation and Development," 21 October 1938, box 14, ABCD 1936–1941. R. Bruce Etheridge to Lewis Dorsett, 11 August 1938, box 16, ABCD 1935–1941.

109. Ray, *Ecology of a Cracker Childhood*, 38.

110. W. G. Wahlenberg, *Longleaf Pine, Its Use, Ecology, Regeneration, Protection, Growth, and Management*; G. Frederick Schwarz, *The Longleaf Pine in Virgin Forest: A Silvicultural Study*, 47–87; W. G. Wahlenberg, "Effects of Fire and Cattle Grazing on Longleaf Pine Lands, as Studied at McNeill, Miss."

111. "In placing the birds regard will be given to protection, cover, food, water, and especial attention will be paid to hunting preserves where conditions are best adapted to the care of the birds." *Whiteville News Reporter*, 1 March 1928. "As to game [on the Bellamy plantation] there are quite a number of pheasants, the original stock of which were placed on the property some two years ago, Mexican quail and deer and wild turkeys in large quantities." Nelson MacRae to Wayne Cunningham, 11 December 1933, box 7, folder 2, Hugh MacRae Papers.

112. J. S. Holmes, "Memorandum to the Director," 13 June 1928; L.D. Coltrane to J. S. Holmes, 5 June 1928, box 3, ADCD 1926–1935. Holmes echoed a concern raised by Eric Norden that "the entire [coast] line of the Eastern part of the State is now being bought up for a very few dollars by wealthy Northerners, and in a few years the natives of North Carolina will be entirely barred from access to the Atlantic Ocean." Norden to Holmes, 28 July 1926, box 1378.32, Eric Norden Collection.

113. *Whiteville News Reporter*, 10 April 1924.

114. *Whiteville News Reporter*, 13 August 1925.

115. Thorndike Saville to J. S. Holmes, n.d., box 8, ABCD 1926–1935.

116. Charles E. Ray to J. S. Holmes, 22 July 1926; K. Clyde Council to J. S. Holmes, 13 July 1926 and 16 July 1926; K. Clyde Council to Governor A. W. McLean, 29 July 1926, box 8, ABCD 1926–1935.

117. John D. Eriksen to R. Bruce Etheridge, 10 March 1938; Etheridge to Eriksen, 22 March 1938, ABCD 1936–1941. *Southport Standard*, 18 August 1904. *Whiteville News Reporter*, 6 August 1925.

118. *Southern Lumberman*, 19 December 1925, contains a 34–page special feature on forestry and reforestation.

119. J. S. Holmes, "Report to the Board of Conservation and Development, Division of Forestry," 22 January 1936, typed manuscript, box 14, ABCD 1936–1941.

120. "Forest Management Section of the Forestry Division," undated typed manuscript (probably 1937), box 14, ABCD 1936–1941.

121. "Activities of the CCC for the period July 1 to December 31, 1937"; also William Maughan, *A Guide to Forestry Activities in North Carolina, South Carolina, and Tennessee*.

122. *Southern Lumberman*, 19 December 1925.

123. *Whiteville News Reporter*, 9 April 1925.

124. R. W. Graeber, "Management, Harvest, and Sale of Pine Timber For Pulpwood," *Circular* 218, 3.

125. Reprinted in *Whiteville News Reporter*, 10 April 1924.

126. Austin Cary to Southeastern Council, 28 December 1933, box 7, folder 2, Hugh MacRae Papers.

Epilogue: "Our Meat from God"

1. Dean, *With Broadax and Firebrand*, 5.

2. See Ted Steinberg, *American Green: The Obsessive Quest for the Perfect Lawn*, 83–104, for an analysis of the environmental disaster that is the golf course.

3. Quoted in William Leach, *Land of Desire: Merchants, Power, and the Rise of a New American Culture*, 390.

4. Wendell Berry, "The Gift of Good Land: Further Essay Cultural and Agricultural," especially 267–281; Aldo Leopold, *A Sand County Almanac*, xviii–xix.

5. Ibid., xviii.

Bibliography

Manuscript Sources

Southern Historical Collection, University of North Carolina–Chapel Hill

Brown Family Papers
Catherina Eliza Clitherall Diaries
Charles Beers Gault Papers
Charles Oscar Beers Papers
John K. Baldwin Diary
Joy and Metcalf Records
Lemuel Bannister Papers
Mercer Family Papers
Pierce & Company Papers
Potter and Platt Family Papers
Thomas David Smith McDowell Papers
William Harris Garland Papers

Rare Book, Manuscript, and Special Collections, Perkins Library, Duke University, Durham, North Carolina

A. J. and William H. Turlington Papers
Alexander Sprunt & Sons Papers
Henry L. Adams Papers
Hogg & Clayton Letterbook
Hugh MacRae Papers
Jacob Mordecai Papers
James Redding Grist Papers
John Huske and James Hogg Papers
John McKay Papers
Oliver H. Jones Papers
Waccamaw Lumber Company Collection
William Walter and Company Papers

Special Collections, J. Y. Joyner Library, East Carolina University, Greenville, North Carolina

Samuel James Rowell Papers

Special Collections, William Madison Randall Library, UNC– Wilmington

High Family Papers
Lake Waccamaw History
Oscar High Papers
Van Eden Papers

Historical Collections, Baker Library, Harvard Business School

R. G. Dun & Company Collection
Forest History Society
Waccamaw Lumber Company Blueprints

Wisconsin Historical Society, University of Wisconsin–Madison

Civilian Conservation Corps Camp Papers

North Carolina State Archives, Raleigh

Brunswick County Population Schedules
Columbus County Population Schedules
Columbus County Records of Deeds
Columbus County Record of Incorporations
Eric Norden Collection
Governors' Letter Books
Bladen County Index to Real Estate Conveyances
James W. Purdie Papers

North Carolina Department of Conservation and Development

Activities of the Board of Conservation and Development (ABCD)
North Carolina Geological and Economic Survey
State Board of Education Swamp Land Records

Published Sources

"C. Vann Woodward's Origins of the New South, 1877–1913: A Fifty-Year Retrospective." *Journal of Southern History* 67, no. 4 (November 2001): 715–826.

"Life and Travel in the Southern States." *Great Republic Monthly* 1 (1859): 80–84.

The Lumberman's Directory of Saw Mills, Shingle Mills and Other Wood Working Factories in the Northwest, South and Southwest, Comprising the States of Michigan, Wisconsin, Minnesota, the Mississippi River from St. Cloud to St. Louis, Delaware, Maryland, Virginia, West Virginia, North Carolina, South Carolina, Georgia, Florida, Alabama, Mississippi, Louisiana, Texas, Kentucky, Tennessee Missouri, Arkansas, Southern Il-

linois, Ohio and Indiana; Arranged by Railroads, Rivers and Seaports. Chicago: W. B. Judson, 1880.

Proceedings of the Southern Reclamation Conference Held in Washington, D. C., December 14 and 15, 1927 under the auspices of the Department of the Interior, Hubert Work, Secretary, and of the Bureau of Reclamation, Elwood Mead, Commissioner. Washington, D.C.: Government Printing Office, 1928.

The Rand McNally Road Atlas. Skokie, Ill.: Rand McNally, 2001.

The Southern Business Directory and General Commercial Advertiser. Vol. 1. Charleston, 1854.

"To Boldly Dump: In Space, Nobody Ever Tidies Up." *The Economist* 29 March 1997: 87–88.

Adams, Herbert B. *The Life and Writings of Jared Sparks: Comprising Selections from His Journals and Correspondence*. Boston: Houghton Mifflin, 1893.

Anderson, Benedict. *Imagined Communities: Reflections on the Origin and Spread of Nationalism*. Rev. ed. London: Verso, 1996.

Andrews, Evangeline Walker, and Charles McLean Andrews, eds. *Journal of a Lady of Quality: Being the Narrative of a Journey from Scotland to the West Indies, North Carolina, and Portugal, in the Years 1774 to 1776*. New Haven: Yale University Press, 1934.

Appleby, Joyce. *Capitalism and a New Social Order: The Republican Vision of the 1790s*. New York: New York University Press, 1984.

Aron, Steve. *How the West Was Lost: The Transformation of Kentucky from Daniel Boone to Henry Clay*. Baltimore: Johns Hopkins University Press, 1996.

Ashe, W. W. *The Forests, Forest Lands, and Forest Products of Eastern North Carolina*. North Carolina Geological Survey Bulletin 5. Raleigh: J. Daniels, 1894.

Associated Committees on Southern Rural Development. *Southern Rural Life, the Problem, the Solution: Opening the Way for Men to Become Farm Owners*. Wilmington, N.C.: Associated Committees on Southern Rural Development, 1929.

Auerbach, Erich. *Mimesis: The Representation of Reality in Western Literature*. Princeton: Princeton University Press, 1953.

Avirett, James Battle. *The Old Plantation: How We Lived in Great House and Cabin before the War*. New York: F. Tennyson Neely, 1901.

Ayers, Edward L. *The Promise of the New South: Life After Reconstruction*. New York: Oxford University Press, 1992.

Bailey, Joseph Cannon. *Seaman Knapp: Schoolmaster of American Agriculture*. New York: Columbia University Press, 1945.

Bailey, Liberty H. *The Country-Life Movement in the United States*. New York: MacMillan, 1911.

Bartram, John. *Diary of a Journey through the Carolinas, Georgia, and Florida from July 1, 1765, to April 10, 1766*. Transactions of the American Philosophical Society 33, Part 1. Philadelphia: American Philosophical Society, 1942.

Bartram, William. *Travels through North and South Carolina, Georgia, East and West Florida*. 1791. Reprint. Charlottesville: University of Virginia Press, 1973.

Battle, Kemp P. *Letters and Documents, Relating to the Early History of the Lower Cape Fear*. James Sprunt Historical Monograph, No. 4. Chapel Hill: University of North Carolina Press, 1903.

Berry, Wendell. *The Gift of Good Land: Further Essay Cultural and Agricultural*. New York: North Point Press, 1981.

———. *The Unsettling of America: Culture & Agriculture*. San Francisco: Sierra Club Books, 1986.

———. *What Are People For?: Essays by Wendell Barry*. New York: North Point Press, 1990.

———. *Sex, Economy, Freedom & Community*. New York: Pantheon, 1993.

Berthoff, Rowland T. "Southern Attitudes toward Immigration, 1865–1914." *Journal of Southern History* 17, no. 3 (August 1951): 328–360.

Bingham, Emily. *Mordecai: An Early American Family*. New York: Hill and Wang, 2003.

Bishop, Nathaniel H. *Voyage of the Paper Canoe: A Geographical Journey of 2500 Miles, from Quebec to the Gulf of Mexico during the Years 1874–5*. Boston: Lee and Shepard, 1878.

Boyd, William K., ed. *William Byrd's Histories of the Dividing Line Betwixt Virginia and North Carolina*. New York: Dover, 1967.

Bourdieu, Pierre. "The Forms of Capital." *Handbook of Theory and Research for the Sociology of Education*. Edited by J. G. Richardson. New York: Greenwood Press, 1986. 214–258.

Bowers, William L. *The Country Life Movement in America, 1900–1920*. Port Washington, N.Y.: Kennikat, 1974.

Boyd, James. "Old Pines." *North Carolina in the Short Story*. Edited by Richard Walser. Chapel Hill: University of North Carolina Press, 1948. 161–181.

Brickell, John. *The Natural History of North Carolina*. 1737. Reprint. Murfreesboro, N.C.: Johnson, 1968.

Brown, Margaret Lynn. *The Wild East: A Biography of the Great Smoky Mountains*. Gainesville: University Press of Florida, 2000.

Bruegel, Martin. *Farm, Shop, Landing: The Rise of a Market Society in the Hudson Valley, 1780–1860*. Durham: Duke University Press, 2002.

Brundage, W. Fitzhugh. *A Socialist Utopia in the New South: The Ruskin Colonies in Tennessee and Georgia, 1894–1901*. Urbana: University of Illinois Press, 1996.

Bushman, Richard. "Family Security in the Transition from Farm to City, 1750–1850." *Journal of Family History* 6 (July 1981): 238–256.

Butler, Carroll B. *Treasures of the Longleaf Pines Naval Stores*. Shalimar, Fla.: Tarkel Publishing, 1998.

Carlton, David L. *Mill and Town in South Carolina, 1880–1920*. Baton Rouge: Louisiana State University Press, 1982.

———. "The Piedmont and Waccamaw Regions: An Economic Comparison." *South Carolina Historical Magazine* 88, no. 2 (1987): 83–100.

Carlton, David L., and Peter A. Coclanis, eds. *Confronting Southern Poverty in the Great Depression: The Report on Economic Conditions of the South with Related Documents.* Boston: Bedford, 1996.

Carstensen, Vernon Rosco. *Farms or Forests: Evolution of a State Land Policy for Northern Wisconsin, 1850–1932.* Madison: University of Wisconsin Press, 1958.

Cashin, Edward J. *William Bartram and the American Revolution on the Southern Frontier.* Columbia: University of South Carolina Press, 2000.

Causey, Gwen. *Great Is Thy Faithfulness: A Glorious Century; A Centennial History, 1899–1999.* Supply, N.C.: Brunswick Baptist Association, 1999.

Cavanaugh, William T. "Killing for the Telephone Company: Why the Nation-State is not the Keeper of the Common Good." *Modern Theology* 20, no. 2 (April 2004): 243–274.

Cayton, Andrew. *The Frontier Republic: Ideology and Politics in the Ohio Country, 1780–1825.* Kent, Ohio: Kent State University Press, 1986.

Cecelski, David. *A Historian's Coast: Adventures into the Tidewater Past.* Winston-Salem: John F. Blair, 2000.

———. *The Waterman's Song: Slavery and Freedom in Maritime North Carolina.* Chapel Hill: University of North Carolina Press, 2001.

Cecelski, David S., and Timothy B. Tyson, eds. *Democracy Betrayed: The Wilmington Race Riot of 1898 and Its Legacy.* Chapel Hill: University of North Carolina Press, 1998.

Censer, Jane Turner. "Southwestern Migration among North Carolina Planter Families: 'The Disposition to Emigrate.'" *Journal of Southern History* 57 (August 1991): 407–426.

Chandler, Alfred D., Jr. "Anthracite Coal and the Beginnings of the Industrial Revolution in the United States." *The Essential Alfred Chandler: Essays toward a Historical Theory of Big Business.* Edited by Thomas K. McCraw. Boston: Harvard Business School Press, 1988. 307–342.

———. *The Visible Hand: The Managerial Revolution in American Business.* Cambridge, Mass.: Belknap, 1977.

Chaplin, Joyce E. "Tidal Rice Cultivation and the Problem of Slavery in South Carolina and Georgia, 1760–1815." *William and Mary Quarterly* 49 (3rd Series), no. 1 (January 1992): 29–61.

———. *An Anxious Pursuit: Agricultural Innovation and Modernity in the Lower South, 1730–1815.* Chapel Hill: University of North Carolina Press, 1993.

Clark, Christopher, ed. "The Transition to Capitalism in America: A Panel Discussion." *The History Teacher* 27, no. 3 (May 1994): 263–288.

Clark, Thomas D. *The Greening of the South: The Recovery of Land and Forest.* Lexington: University Press of Kentucky, 1984.

Clark, Walter, ed. *The State Records of North Carolina*, vols. 11–26. Goldsboro, N.C.: Nash Brothers, 1905.

Clepper, Henry. *Professional Forestry in the United States*. Baltimore: John Hopkins Press, 1971.

Clifton, James M. "Golden Grains of White: Rice Planting on the Lower Cape Fear." *North Carolina Historical Review* 50, no. 4 (October 1973): 365–393.

Coca-Cola Bottling Company. *Our America. Lumber: Trees, Timber, Wood Products*. Atlanta, 1943.

Cochran, Thomas C. *Frontiers of Change: Early Industrial Development in the United States and Britain*. New York: Oxford University Press, 1981.

Coclanis, Peter A. "Urbs in Horto." *Reviews in American History* 20 (1992): 14–20.

Conkin, Paul K. *Tomorrow a New World: The New Deal Community Program*. Ithaca, N.Y.: Cornell University Press, 1959.

Corbin, David Allen. *Life, Work and Rebellion in the Coal Fields: The Southern West Virginia Miners, 1880–1922*. Urbana: University of Illinois Press, 1981.

Corey, John Faris. "The Colonization and Contributions of Emmigrants Brought to Southeastern North Carolina by Hugh MacRae." Master's thesis, Appalachian State Teachers College (Boone, N.C.), 1957.

Cowdrey, Albert E. *This Land, This South: An Environmental History*. Rev. ed. Lexington: University Press of Kentucky, 1996.

Crawford, Clifton E. *Bladen County History: Information Pertaining to the History and Development of Bladen County taken from the files of the Bladen Journal, Elizabethtown, North Carolina*. 1950.

Cronon, William. *Nature's Metropolis: Chicago and the Great West*. New York: W. W. Norton, 1991.

———. "Kennecott Journey: The Paths out of Town." *Under an Open Sky: Rethinking America's Western Past*. Edited by Cronon, George Miles, and Jay Gitlin. New York: W. W. Norton, 1992. 28–51.

———."The Trouble with Wilderness; or, Getting Back to the Wrong Nature." *Uncommon Ground: Rethinking the Human Place in Nature*. Edited by William Cronon. New York: W. W. Norton, 1996. 69–90.

———. "The Uses of Environmental History." *Environmental History Review* (Fall 1993): 1–22.

Crow, Jeffrey J. "Slave Rebelliousness and Social Conflict in North Carolina, 1775 to 1802." *William and Mary Quarterly* 37, no. 1 (January 1980): 79–102.

Curtin, Philip D. *The Rise and Fall of the Plantation Complex: Essays in Atlantic History*. 2nd ed. Cambridge: Cambridge University Press, 1998.

Curtis, Walter Gilman. *Reminiscences of Wilmington and Smithville-Southport, 1848–1900*. Southport: *Southport Herald*, 1905.

Danbom, David B. *The Resisted Revolution: Urban America and the Industrialization of Agriculture, 1900–1930*. Ames: Iowa State University Press, 1979.

Dean, Warren. *With Broadax and Firebrand: The Destruction of the Brazilian Atlantic Forest.* Berkeley: University of California Press, 1995.

———. "The Tasks of Latin American Environmental History." *Changing Tropical Forests: Historical Perspectives on Today's Challenges in Central and South America.* Edited by Harold K. Steen and Richard P. Tucker. Durham: Forest History Society, 1992. 5–15.

DeCanio, Stephen. *Agriculture in the Postbellum South: The Economics of Production and Supply.* Cambridge: MIT Press, 1974.

Defebaugh, James Elliott. *History of the Lumber Industry of America.* 2 vols. Chicago: American Lumberman, 1906–1907.

Denoon, Donald, and Philippa Mein-Smith, with Marivic Wyndham. *A History of Australia, New Zealand and the Pacific.* Malden, Mass.: Blackwell, 2000.

De Vries, Hugo. *Naar Californië II: Reisherinneringen.* Haarlem, The Netherlands: Tjeenk Willink, 1907.

Dial, Adolph L., and David K. Eliades. *The Only Land I Know: A History of the Lumbee Indians.* Syracuse: Syracuse University Press, 1996.

Dowd, Douglas. *Capitalism and Its Economics: A Critical History.* London: Pluto, 2000.

Doyle, Don H. *Social Order of a Frontier Community: Jacksonville, Illinois, 1825–1870.* Urbana: University of Illinois Press, 1978.

Drobney, Jeffrey A. *Lumbermen and Log Sawyers: Life, Labor and Culture in the North Florida Timber Industry, 1830–1930.* Macon, Ga.: Mercer University Press, 1997.

Dunaway, Wilma A. *The First American Frontier: Transition to Capitalism in Southern Appalachia, 1700–1860.* Chapel Hill: University of North Carolina Press, 1996.

Dunn, Richard. *Sugar and Slaves: The Rise of the Planter Class in the English West Indies, 1624–1713.* Chapel Hill: University of North Carolina Press, 1972.

Dusinberre, William. *Them Dark Days: Slavery in the American Rice Swamps.* New York: Oxford University Press, 1996.

Earley, Lawrence S. *Looking for Longleaf: The Fall and Rise of an American Forest.* Chapel Hill: University of North Carolina Press, 2004.

Edmunds, Pocahontas Wight. *Tales of the North Carolina Coast.* Raleigh: Edwards & Broughton, 1986.

Ekirch, A. Roger. *"Poor Carolina": Politics and Society in Colonial North Carolina, 1729–1776.* Chapel Hill: University of North Carolina Press, 1981.

Emmerich, Susan Drake. "Faith-Based Stewardship and Resolution of Environmental Conflict: An Ethnography of an Action Research Case of Tangier Island Watermen in the Chesapeake Bay." PhD diss., University of Wisconsin–Madison, 2003.

Escott, Paul D. *Many Excellent People: Power and Privilege in North Carolina, 1850–1900.* Chapel Hill: University of North Carolina Press, 1985.

Evans, W. McKee. *To Die Game: The Story of the Lowry Band, Indian Guerrillas of Reconstruction.* Baton Rouge: Louisiana State University Press, 1971.

Facey, Douglas E., and Michael J. Van Den Avyle. *Species Profiles: Life Histories and Environmental Requirements of Coastal Fishes and Invertebrates (North Atlantic): American*

Eel. Biological Report 82. Washington, D.C.: Department of the Interior, Coastal U.S. Fish and Wildlife Service, 1987.

Faragher, John Mack. *Sugar Creek: Life on the Illinois Prairie*. New Haven: Yale University Press, 1986.

Fickle, James E. *The New South and the "New Competition": Trade Association Development in the Southern Pine Industry*. Urbana: University of Illinois Press, 1980.

Fields, Barbara Jeanne. *Slavery and Freedom on the Middle Ground: Maryland in the Nineteenth Century*. New Haven: Yale University Press, 1985.

Fite, Gilbert. *Cotton Fields No More: Southern Agriculture, 1865–1980*. Lexington: University Press of Kentucky, 1984.

Ford, Lacy K., Jr. *Origins of Southern Radicalism: The South Carolina Upcountry, 1800–1860*. New York: Oxford University Press, 1988.

Frankenberg, Dirk. *The Nature of North Carolina's Southern Coast: Barrier Islands, Coastal Waters, and Wetlands*. Chapel Hill: University of North Carolina Press, 1997.

Friedman, Milton. *Capitalism and Freedom*. Chicago: University of Chicago Press, 1962.

Gaston, Paul M. *The New South Creed: A Study in Southern Mythmaking*. New York: Knopf, 1970.

Gates, Paul Wallace. "Federal Land Policy in the South, 1866–1888." *Journal of Southern History* 6, no. 3 (August 1940): 303–330.

Giarchi, George Giacinto. "Caught in the Nets: A Critical Examination of the Use of the Concept of 'Networks' in Community Development Studies." *Community Development Journal* 36, no. 1 (January 2001): 63–71.

Gilje, Paul A. "The Rise of Capitalism in the Early Republic." *Wages of Independence: Capitalism in the Early American Republic*. Edited by Paul A. Gilje. Madison, Wis.: Madison House, 1997. 1–22.

Girl Scouts of the United States of America. *Girl Scout Handbook*. Rev. ed. New York: Girl Scouts Inc., 1929.

Glaab, Charles Nelson. *Kansas City and the Railroads: Community Policy in the Growth of a Regional Metropolis*. Lawrence: University Press of Kansas, 1993.

Glacken, Clarence J. *Traces on the Rhodian Shore: Nature and Culture in Western Thought from Ancient Times to the End of the Eighteenth Century*. Berkeley: University of California Press, 1967.

Goddard, Frederick B. *Where to Emigrate and Why: Homes and Fortunes in the Boundless West and the Sunny South; Their Climate, Scenery, Soil, Productions, Railroads, Mining Interests, and General Resources; the Cost of Farm Lands, How to Obtain Titles, the Homestead and Other Land Laws, The Rates of Wages, &c., &c*. Philadelphia: The People's Publishing Company, 1869.

Goodwin, O. C. *Eight Decades of Forestry Firsts: A History of Forestry in North Carolina, 1889–1969*. Raleigh: North Carolina Forest Service, 1969.

Gough, Robert. *Farming the Cutover: A Social History of Northern Wisconsin, 1900–1940*. Lawrence: University of Kansas Press, 1997.

Graeber, R. W. *Management, Harvest, And Sale of Pine Timber For Pulpwood*. North Carolina State College of Agriculture and Engineering Extension Circular 218. Raleigh, 1937.

Grantham, Dewey. *Southern Progressivism: The Reconciliation of Progress and Tradition*. Knoxville: University of Tennessee Press, 1983.

Grey, Zane. *The Young Forester*. New York: Grosset & Dunlap, 1910.

Grootaert, Christiaan, and Thierry van Bastelaer. "Social Capital: From Definition to Measurement." *Understanding and Measuring Social Capital: A Multidisciplinary Tool for Practitioners*. Edited by Grootaert and van Bastelaer. Washington, D.C.: The World Bank, 2002. 1–15.

Grove, Richard. *Green Imperialism: Colonial Expansion, Tropical Island Edens, and the Origins of Environmentalism, 1600–1860*. Cambridge, United Kingdom: Cambridge University Press, 1995.

Gwartney, James D., and Richard L. Stroup. *Economics: Private and Public Choice*. 7th ed. Fort Worth, Tex.: Dryden Press, 1995.

Haggard, Ted. *Dog Training, Fly Fishing, & Sharing Christ in the 21st Century*. Nashville: Nelson Books, 2002.

Hahn, Steven, and Jonathan Prude, eds. *The Countryside in the Age of Capitalist Transformation: Essays in the Social History of Rural America*. Chapel Hill: University of North Carolina Press, 1985.

Hall, Basil. *Travels in North America, in the Years 1827 and 1828*. New York: Arno Press, 1974.

Hall, David D. *Worlds of Wonder, Days of Judgment: Popular Religious Belief in New England*. New York: Knopf, 1989.

Hall, Jacqueline Dowd, et al. *Like a Family: The Making of a Southern Cotton Mill World*. Chapel Hill: University of North Carolina Press, 1987.

Halttunen, Karen. *Confidence Men and Painted Women: A Study of Middle-Class Culture in America, 1830–1870*. New Haven: Yale University Press, 1982.

Hammond, John Martin. *Winter Journeys in the South*. Philadelphia: Lippincott, 1916.

Hanlon, Howard A. *The Bull Hunchers: A Saga of the Three and a Half Centuries of Harvesting the Forest Crops of the Tidewater Low Country*. Parsons, W.Va.: McClain, 1970.

Hart, Albert Bushnell. *The Southern South*. 1910. New York: Negro Universities Press, 1969.

Haynes, William. "Through a Jungle to the Old South." *Outing* 65 (January 1915): 478–486.

———. "Down the Old Lumbee." *Field and Stream* 23 (May 1918): 5–7, 48–49.

Hodgson, Adam. *Letters from North America, Written during a Tour in the United States and Canada*, Vol. 1. London, 1824.

Hays, Samuel P. *Conservation and the Gospel of Efficiency: The Progressive Conservation Movement, 1890–1920*. Cambridge: Harvard University Press, 1959.

———. *Explorations in Environmental History: Essays*. Pittsburgh: University of Pittsburgh Press, 1998.

Hearden, Patrick J. *Independence and Empire: The New South's Cotton Mill Campaign*. DeKalb: Northern Illinois University Press, 1982.

Helgeson, Alan. *Farms in the Cutover: Agricultural Settlement in Northern Wisconsin*. Madison: State Historical Society of Wisconsin, 1962.

Hendricks, W. C., ed. *Bundle of Troubles and Other Tarheel Tales, by Workers of the Writers' Program of the Work Projects Administration in the State of North Carolina*. Durham: Duke University Press, 1943.

Henry Disston & Sons. *The Saw in History: A Comprehensive Description of the Development of This Most Useful of Tools from the Earliest Times to the Present Day*. 2nd ed. Philadelphia: Disston & Sons, 1916.

Higgs, Robert. *Competition and Coercion: Blacks in the American Economy, 1865–1914*. Cambridge: Cambridge University Press, 1977.

Holloway, Thomas H. *Immigrants on the Land: Coffee and Society in São Paulo, 1886–1934*. Chapel Hill: University of North Carolina Press, 1980.

Huggins, Nathan Irvin. *Harlem Renaissance*. New York: Oxford University Press, 1971.

Inchausti, Robert. *Subversive Orthodoxy: Outlaws, Subversives, and Other Christians in Disguise*. Grand Rapids, Mich.: Brazos Press, 2005.

International Workers of the World. *The Lumber Industry and Its Workers*. 2nd ed. Chicago: IWW Publishing Bureau, 1920.

Jacoby, Karl. *Crimes against Nature: Squatters, Poachers, Thieves, and the Hidden History of American Conservation*. Berkeley: University of California Press, 2003.

Jensen, Vernon H. *Lumber and Labor*. New York: Arno, 1971.

John, Richard R. "Elaborations, Revisions, Dissents: Alfred D. Chandler, Jr.'s *The Visible Hand* after Twenty Years." *Business History Review* 71 (Summer 1997): 151–200.

Jones, Jacqueline. *The Dispossessed: America's Underclasses from the Civil War to the Present*. New York: Basic Books, 1992.

Jones, Thomas H. *The Experience of Thomas H. Jones, Who Was a Slave for Forty-three Years: Written by a Friend, as Related to Him by Brother Jones*. Boston: Bazin & Chandler, 1862.

Kates, James. *Planning a Wilderness: Regenerating the Great Lakes Cutover Region*. Minneapolis: University of Minnesota Press, 2001.

Keith, Jeanette. *Country People in the New South: Tennessee's Upper Cumberland*. Chapel Hill: University of North Carolina Press, 1995.

Kellar, Herbert Anthony, ed. *Solon Robinson, Pioneer and Agriculturist: Selected Writings*. Vol. 2, *1846–1851*. Indianapolis: Indiana Historical Bureau, 1936.

Kemble, Frances Anne. *Records of Later Life*. Vol. 1. London, 1882.

Kirke, Edmund [James R. Gilmore, pseud.]. *Among the Pines: Or, South in Secession-Time*. New York: J. R. Gilmore, 1862.

Kirby, Jack Temple. *Poquosin: A Study of Rural Landscape and Society*. Chapel Hill: University of North Carolina Press, 1995.

———. "Clarence Poe's Vision of a Segregated 'Great Rural Civilization.'" *South Atlantic Quarterly* 68, no. 1 (1969): 27–38.

———. *Darkness at the Dawning: Race and Reform in the Progressive South*. Philadelphia: Lippincott, 1972.

———. *Mockingbird Song: Ecological Landscapes of the South*. Chapel Hill: University of North Carolina Press, 2006.

Klein, Mary. "Coming Full Circle: The Study of Big Business since 1950." *Enterprise & Society* 2, no. 3 (September 2001): 425–460.

Knight, Franklin W., and Peggy K. Liss, eds. *Atlantic Port Cities: Economy, Culture, and Society in the Atlantic World, 1650–1850*. Knoxville: The University of Tennessee Press, 1991.

Kologiski, Russell L. *The Phytosociology of the Green Swamp, North Carolina*. North Carolina Agricultural Experiment Station Technical Bulletin 250. Raleigh: North Carolina Agricultural Experiment Station, 1977.

Kusel, Jonathan, and Elisa Adler, eds. *Forest Communities, Community Forests*. Lanham, Md.: Rowman & Littlefield, 2003.

Kuyper, Abraham. *Lectures on Calvinism*. Grand Rapids: Eerdmans, 2002.

Leach, William. *Land of Desire: Merchants, Power, and the Rise of a New American Culture*. New York: Vintage, 1994.

LeBlond, Richard. *Inventory of the Natural Areas and Rare Species of Brunswick County, North Carolina*. Raleigh: Department of Environment, Health, and Natural Resources, 1995.

———. *Inventory of the Natural Areas and Rare Species of Columbus County, North Carolina*. Raleigh: Department of Environment, Health, and Natural Resources, 1995.

Leonard, S. *Inventory of Populations of* Thalictrum cooleyi *and Its Occurrence Sites in North Carolina: Report to the North Carolina Natural Heritage Program*. Raleigh: North Carolina Natural Heritage Program, 1987.

Leopold, Aldo. *A Sand County Almanac*. New York: Ballantine, 1970.

Lewis, Ronald L. *Transforming the Appalachian Countryside: Railroads, Deforestation, and Social Change in West Virginia, 1880–1920*. Chapel Hill: University of North Carolina Press, 1998.

———. *Black Coal Miners in America: Race, Class, and Community Conflict, 1780–1980*. Lexington: University Press of Kentucky, 1987.

Lin, Nan. *Social Capital: A Theory of Social Structure and Action*. Cambridge: Cambridge University Press, 2001.

Link, William A. *The Paradox of Southern Progressivism, 1880–1930*. Chapel Hill: University of North Carolina Press, 1992.

Little, Ann Courtney Ward, ed. *Columbus County, North Carolina: Recollections and Records*. Whiteville: Columbus County Bicentennial Commission, 1976.

Litwack, Leon F. *Trouble in Mind: Black Southerners in the Age of Jim Crow*. New York: Knopf, 1998.

Lumbermen's Credit Association. *The Lumbermen's Credit Rating Book: Clancy's Red*

Book and Lumbermen's Blue Book of the Lumbermen's Red and Blue Book Service. 112th ed. Chicago: Lumbermen's Credit Association, 1937.

MacIntyre, Alasdair. *After Virtue.* 2nd ed. Notre Dame, Ind.: University of Notre Dame Press, 1984.

Malinowski, Bronislaw. *Myth in Primitive Psychology.* New York: W. W. Norton, 1926.

Manguel, Alberto. *Into the Looking-Glass Wood: Essays on Books, Reading, and the World.* San Diego: Harcourt, 1998.

Marty, Rebecca S. "The Floodplain Plant Communities of a Blackwater River: The Waccamaw River, North Carolina." Master's thesis, Duke University, 1988.

Marx, Leo. *The Machine in the Garden: Technology and the Pastoral Ideal in America.* New York: Oxford University Press, 1964.

Maughan, William. *A Guide to Forestry Activities in North Carolina, South Carolina, and Tennessee.* Washington, D. C.: Society of American Foresters, 1939.

Maull, Allen. *Eastern North Carolina for the Farmer.* Wilmington: Atlantic Coast Line Passenger Department, 1916.

Maxwell, Robert S., and Robert D. Baker. *Sawdust Empire: The Texas Lumber Industry, 1830–1940.* College Station: Texas A&M University Press, 1983.

McCally, David. *The Everglades: An Environmental History.* Gainesville: University Press of Florida, 1999.

McDonald, Roderick A. *The Economy and Material Culture of Slaves: Goods and Chattels on the Sugar Plantations of Jamaica and Louisiana.* Baton Rouge: Louisiana State University Press, 1993.

McLaurin, Melton Alonza. *Paternalism and Protest: Southern Cotton Mill Workers and Organized Labor, 1875–1905.* Westport, Conn.: Greenwood Publishing Group, 1971.

Melville, Elinor G. K. *A Plague of Sheep: Environmental Consequences of the Conquest of Mexico.* Cambridge: Cambridge University Press, 1997.

———. "Environmental and Social Change in the Valle del Mezquital, Mexico, 1521–1600." *Comparative Studies in Society and History* 32, no. 1 (January 1990): 24–53.

Merchant, Carolyn. *The Death of Nature: Women, Ecology and the Scientific Revolution.* San Francisco: Harper Collins, 1990.

Merrill, Michael. "Putting 'Capitalism' in Its Place: A Review of Recent Literature." *William and Mary Quarterly* 52, no. 2 (April 1995): 315–326.

Messina, Michael G., and William H. Conner, eds. *Southern Forested Wetlands: Ecology and Management.* Boca Raton, Fla.: Lewis, 1998.

Michael, Marvin L., and Lorin Lee Cary. *Slavery in North Carolina, 1748–1775.* Chapel Hill: University of North Carolina Press, 1995.

Miller, Randall Martin. *The Cotton Mill Movement in Antebellum Alabama.* New York: Arno Press, 1978.

Miller, Shawn William. *Fruitless Trees: Portuguese Conservation and Brazil's Colonial Timber.* Palo Alto: Stanford University Press, 2000.

Mooijweer, Marianne. *De Amerikaanse Droom van Frederik van Eeden*. Amsterdam: Bataafsche Leeuw, 1996.

Murray, Amelia M. *Letters from the United States, Cuba, and Canada*. New York: Putnam, 1856.

Musselwhite, Eugene McMillan. *The "Iron Horse" Comes to Robeson, a County of a "Rip Van Winkle State."* Lumberton, N. C., 1968.

Nash, Roderick. Nash, *Wilderness and the American Mind*. 3rd ed. New Haven: Yale University Press, 1982.

Needham, Andrew. "'Qualified to Succeed': Elwood Mead and the Durham Agricultural Colony, 1915–1919." *Ex Post Facto* 5 (1996).

Nelson, Connie, and Floyd Harris. *The Film Junkie's Guide to North Carolina*. Winston-Salem: John F. Blair, 2004.

Newby, I. A. *Plain Folk in the New South: Social Change and Cultural Persistence, 1880–1915*. Baton Rouge: Louisiana State University Press, 1989.

North Carolina Department of Agriculture. *North Carolina and Its Resources*. Winston, N.C.: Stewart, 1896.

North Carolina Department of Conservation and Development. *Hunting in North Carolina*. Bulletin 36. Raleigh, 1928.

North Carolina Division of Environmental Management. *A Field Guide to North Carolina Wetlands: Report No. 96–01*. Raleigh: North Carolina Department of Environment, Health, and Natural Resources, 1996.

North Carolina. Geological and Economic Survey. *Proceedings of the Second Annual Convention of North Carolina Forestry Association Held at Raleigh, North Carolina, February 21, 1912*. Economic Paper 25. Raleigh: Edwards & Broughton, 1912.

———. *Organization of Co-Operative Forest-Fire Protective Areas in North Carolina, Being the Proceedings of the Special Conference on Forest Fire Protection Held as Part of the Conference on Forestry and Nature Study, Montreat, N. C., July 8, 1915*. Economic Paper 42. Raleigh: E. M. Uzzell, 1915.

North Carolina Land Company. *A Guide to Capitalists and Emigrants: Being a Statistical and Descriptive Account of the Several Counties of the State of North Carolina, United States of America; Together with Letters of Prominent Citizens of the State in Relation to the Soil, Climate, Productions, Minerals, &c., and an Account of the Swamp Lands of the State*. Raleigh: Nichols and Gorman, 1869.

North Carolina Literary Board. *Swamp Lands of the State of North Carolina: Facts for Emigrants and Capitalists*. Raleigh: W. E. Pell, 1867.

North Carolina Pine Association. *Directory of Pine Lumber Manufacturers in Virginia, North Carolina, South Carolina, Georgia and Florida*. Norfolk: The North Carolina Pine Association, 1907.

Novak, Barbara. *Nature and Culture: American Landscape and Painting, 1825–1875*. New York: Oxford University Press, 1980.

Oden, Jack Porter. "Development of the Southern Pulp and Paper Industry, 1900–1970." PhD diss., Mississippi State University, 1973.

Odum, Howard W. *Southern Regions of the United States*. Chapel Hill: University of North Carolina Press, 1936.

Oelschlaeger, Max. *The Idea of Wilderness: From Prehistory to the Age of Ecology*. New Haven: Yale University Press, 1991.

Olmsted, Frederick Law. *A Journey in the Seaboard Slave States, in the Years 1853–1854, with Remarks on Their Economy*. 1856. New York: Putnam, 1904.

Olson, Sherry. *The Depletion Myth: A History of Railroad Use of Timber*. Cambridge: Harvard University Press, 1971.

Oshinsky, David A. *Worse than Slavery: Parchman Farm and the Ordeal of Jim Crow Justice*. New York: Free Press, 1996.

Outland, Robert B., III. "Slavery, Work, and the Geography of the North Carolina Naval Stores Industry, 1835–1860." *Journal of Southern History* 62, no. 1 (February 1996): 27–56.

———. *Tapping the Pines: The Naval Stores Industry in the American South*. Baton Rouge: Louisiana State University Press, 2004.

———. "Another New South: Patterns of Continuity in the Southern Naval Stores Industry." PhD diss., Louisiana State University, 1999.

Pace, Antonio, transl. and ed. *Luigi Castiglioni's Viaggio: Travels in the United States of North America, 1785–87*. Syracuse: Syracuse University Press, 1983.

Parker, V. Thomas. "The Community of an Individual: Implications for the Community Concept." *OIKOS* 104 (2004): 24–34.

Pawson, Eric, and Tom Brooking, eds. *Environmental Histories of New Zealand*. Melbourne: Oxford University Press, 2002.

Perry, Percival. "The Naval-Stores Industry in the Old South, 1790–1860." *Journal of Southern History* 34, no. 4 (November 1968): 509–526.

Phillips, Catharine. *Memoirs of the Life of Catharine Phillips, To Which Are Added Some of Her Epistles*. London: J. Phillips and Son, 1797.

Phillips, Ulrich B. *A History of Transportation in the Eastern Cotton Belt to 1860*. New York: Columbia University Press, 1908.

Pinchot, Gifford, and W. W. Ashe. *Timber Trees and Forests of North Carolina*. North Carolina Geological Survey Bulletin 6. Winston: Stewart, 1897.

Pope-Hennessy, Una, ed. *The Aristocratic Journey: Being the Outspoken Letters of Mrs. Basil Hall, Written During a Fourteen Months' Sojourn in America, 1827–1828*. New York: Putnam, 1931.

Prather, H. Leon. *We Have Taken a City: Wilmington Racial Massacre and Coup of 1898*. Rutherford, N.J.: Fairleigh Dickinson University Press, 1984.

Pruitt, Bettye Hobbs. "Self-Sufficiency and the Agricultural Economy." *William and Mary Quarterly* 41, no. 3 (July 1984): 333–364.

Putnam, Robert. *Bowling Alone: The Collapse and Revival of American Community*. New York: Simon & Schuster, 2000.

Pyne, Stephen J. *Fire in America: A Cultural History of Wildland and Rural Fire*. Princeton: Princeton University Press, 1982.

Rand, Ayn. *Capitalism: The Unknown Ideal*. New York: Signet, 1967.

Rand, McNally & Company. *Directory and Shipping Guide of Lumber Mills and Lumber Dealers in United States and Canada, Containing a Complete List of Saw Mills, Shingle Mills, Stave Mills, Planing Mills, Sash, Door and Blind Factories, and Dealers in Lumber, Together with the Laws for Material-Men, and a Digest of the Statutes Affecting Lumbermen and Lumbering. Also the Railway and Express Companies to Every Town or City wherein Above Interests Are Located*. Chicago: Rand, McNally & Company, 1884.

Ransom, Roger L., and Richard Sutch. *One Kind of Freedom: The Economic Consequences of Emancipation*. Cambridge: Cambridge University Press, 1977.

Raver, Anne. "Human Nature: A Georgia Daughter's Lullaby of the Pines." *New York Times*, 27 April 2000.

Ray, Janisse. *Ecology of a Cracker Childhood*. Minneapolis: Milkweed, 1999.

Reaves, Bill. *Southport (Smithville): A Chronology*. Vol. 2. Southport: Southport Historical Society, 1990.

Reich, Steven Andrew. "The Making of a Southern Sawmill World: Race, Class, and Rural Transformation in the Piney Woods of East Texas, 1830–1930." PhD diss., Northwestern University, 1998.

Reidy, Joseph P. *From Slavery to Agrarian Capitalism in the Cotton Plantation South: Central Georgia, 1800–1880*. Chapel Hill: University of North Carolina Press, 1992.

Reiger, John F. *American Sportsmen and the Origins of Conservation*. New York: Winchester, 1975.

———. "Wildlife, Conservation, and the First Forest Reserve." *American Forests: Nature, Culture, and Politics*. Edited by Char Miller. Lawrence: University Press of Kansas, 1997. 35–47.

Rights, Douglas L. *The American Indian in North Carolina*. 2nd ed. New York: John F. Blair, 1988.

Robeson, Elizabeth Ellis. *The Diary of Elizabeth Ellis Robeson: Bladen County, North Carolina, from 1847 to 1866*. 1970.

Rome, Abigail. "Vegetation Variation in a Pine-Wiregrass Savanna in the Green Swamp (North Carolina)." Master's thesis, Duke University, 1986.

Roper, Moses. *A Narrative of the Adventures and Escape of Moses Roper, from American Slavery: With a Preface by the Rev. T. Price, D. D.* Philadelphia: Merrihew & Gunn, 1838.

Rose, Anne C. *Voices of the Marketplace: American Thought and Culture, 1830–1860*. New York: Twayne, 1995.

Rosen, Christine Meisner, and Christopher C. Sellers. "The Nature of the Firm: Towards an Ecocultural History of Business." *Business History Review* 73, no. 4 (Winter 1999): 577–600.

Ross, Thomas E. *American Indians in North Carolina: Geographic Interpretations*. Southern Pines, N.C.: Karo Hollow Press, 1999.

Rothenberg, Winifred Barr. *From Market-Places to a Market Economy: The Transfor-*

mation of Rural Massachusetts, 1750–1850. Chicago: University of Chicago Press, 1992.

Sale, Kirkpatrick. *The Green Revolution: The American Environmental Movement, 1962–1992*. New York: Hill and Wang, 1993.

Sanderson, Eric W., et al."The Human Footprint and the Last of the Wild." *BioScience* 52, no. 10 (October 2002): 891–904.

Sargent, Charles S. *Report on the Forests of North America (Exclusive of Mexico)*. Washington, D.C.: U.S. Government Printing Office, 1884.

Saunders, William Laurence, ed. *The Colonial Records of North Carolina*. Vols. 1–10. Raleigh: P. M. Hale, 1886–1890.

Saxenian, AnnaLee. *Regional Advantage: Culture and Competition in Silicon Valley and Route 128*. Cambridge: Harvard University Press, 1994.

Schafale, Michael P. *Inventory of Longleaf Pine Natural Communities in North Carolina*. Raleigh: North Carolina Natural Heritage Program, 1994.

Schama, Simon. *Landscape and Memory*. New York: Vintage, 1995.

Schneider, Richard Eston. "The Effect of Variation in Season of Burning on a Pine-Wiregrass Savanna in the Green Swamp, North Carolina." PhD diss., Duke University, 1988.

Schulman, Bruce J. *The Seventies: The Great Shift in American Culture, Society, and Politics*. Cambridge, Mass.: Da Capo, 2001.

Schwartz, Stuart B. *Sugar Plantations in the Formation of Brazilian Society: Bahia, 1550–1835*. Cambridge: Cambridge University Press, 1985.

Schwarz, G. Frederick. *The Longleaf Pine in Virgin Forest: A Silvicultural Study*. New York: Wiley & Sons, 1907.

Schweid, Richard. *Consider the Eel*. Chapel Hill: University of North Carolina Press, 2002.

Scobie, James R. *Revolution on the Pampas: A Social History of Argentine Wheat, 1860–1910*. Austin: University of Texas Press, 1964.

Scott, James C. *Seeing Like a State: How Certain Schemes to Improve the Human Condition Have Failed*. New Haven: Yale University Press, 1998.

———. *The Moral Economy of the Peasant: Rebellion and Subsistence in Southeast Asia*. New Haven: Yale University Press, 1976.

Scranton, Philip. *Endless Novelty: Specialty Production and American Industrialization, 1865–1925*. Princeton: Princeton University Press, 1997.

Sellers, Charles Grier. *The Market Revolution: Jacksonian America, 1815–1846*. New York: Oxford University Press, 1991.

Sharlet, Jeff. "Inside America's Most Powerful Megachurch." *Harper's Magazine*, May 2005, 41–54.

Sheriff, Carol. *The Artificial River: The Erie Canal and the Paradox of Progress, 1817–1862*. New York: Hill and Wang, 1996.

Sider, Gerald. *Living Indian Histories: The Lumbee and Tuscarora People in North Carolina*. Chapel Hill: University of North Carolina Press, 2006.

Silver, Timothy. *A New Face on the Countryside: Indians, Colonists, and Slaves in South Atlantic Forests, 1500–1800*. Cambridge: Cambridge University Press, 1990.

Simpson, Bland. *Into the Sound Country: A Carolinian's Coastal Plain*. Chapel Hill: University of North Carolina Press, 1997.

Siry, Joseph V. *Marshes of the Ocean Shore: Development of an Ecological Ethic*. College Station: Texas A&M University Press, 1984.

Sitton, Thad, and James H. Conrad. *Nameless Towns: Texas Sawmill Communities, 1880–1942*. Austin: University of Texas Press, 1998.

Skidmore, Thomas E. *Black into White: Race and Nationality in Brazilian Thought*. New York: Oxford University Press, 1974.

Slaughter, Thomas P. *The Natures of John and William Bartram: Two Pioneering Naturalists, Father and Son, in the Wilderness of Eighteenth-Century America*. New York: Vintage, 1996.

Southern Immigration Association of America. *Proceedings of the Annual Session of the Southern Immigration Association of America*. Nashville: Howell, 1884.

Southern Land Congress. *Soldier Settlements in the South: Being the Full and Complete Report of the Southern Land Congress*. Baltimore, 1918.

Southern Pine Association. *Lumber Awakes! Official Report of the First Annual Meeting of the Subscribers to the Southern Pine Association Held at the Grunewald Hotel, New Orleans, Feb. 23–24, 1916*. New Orleans: Southern Pine Association, 1916.

———. *Southern Pine and Reconstruction: Official Report of the Fourth Annual Meeting of the Subscribers to the Southern Pine Association Held at Grunewald Hotel, New Orleans, Feb. 25–26, 1919*. New Orleans: Southern Pine Association, 1919.

Sprunt, James L. *The Story of Orton Plantation, with References to Other Places of Interest on the Cape Fear River, as Gathered from Various Histories and Documents*. Rev. ed. Wilmington, 1966.

———. *Chronicles of the Cape Fear River, 1660–1916*. 2nd ed. Raleigh: Edwards & Broughton, 1916.

Stacey, Margaret. "The Myth of Community Studies." *The British Journal of Sociology* 20, no. 2 (June 1969): 134–147.

Starobin, Robert S. "The Economics of Industrial Slavery in the Old South." *Economics, Industrialization, Urbanization, and Slavery*. Edited by Paul Finkelman. New York: Garland, 1989. 483–526.

Steinberg, Ted. *American Green: The Obsessive Quest for the Perfect Lawn*. New York: W. W. Norton, 2007.

Stewart, Mart A. *"What Nature Suffers to Groe": Life, Labor, and Landscape on the Georgia Coast, 1680–1920*. Athens: University of Georgia Press, 1996.

Stover, John F. *The Railroads of the South, 1865–1900: A Study in Finance and Control*. Chapel Hill: University of North Carolina Press, 1955.

Strickland, John Scott. "Traditional Culture and Moral Economy: Social and Economic Change in the South Carolina Low Country, 1865–1910." *The Countryside in the Age of Capitalist Transformation: Essays in the Social History of Rural America*. Edited by

Steven Hahn and Jonathan Prude. Chapel Hill: University of North Carolina Press, 1985. 141–178.

Stride, Robert. *Southport, North Carolina, and What the Completion of the Wilmington, Brunswick & Southport Railroad Means toward Its Future Development*. 2nd ed. Wilmington, 1913.

Strole, Glenn F. *Chadbourn and her Sunny South Colony: A Narrative History*. Rev. ed. Whiteville: The *News Reporter Company*, 1976.

Taylor, Alan. *Liberty Men and Great Proprietors: The Revolutionary Settlement on the Maine Frontier, 1760–1820*. Chapel Hill: University of North Carolina Press, 1990.

Taylor, George Rogers. *The Transportation Revolution, 1815–1860*. New York: Rinehart, 1951.

Taylor, Joseph E., III. *Making Salmon: An Environmental History of the Northwest Fisheries Crisis*. Seattle: University of Washington Press, 1999.

Taylor, Rosser Howard. "Slaveholding in North Carolina: An Economic View." *James Sprunt Historical Publications* 18, nos. 1–2 (1926): 38–40.

Thompson, E. P. "The Moral Economy of the English Crowd in the Eighteenth Century." *Past and Present* 50 (1971): 76–136.

———. "The Moral Economy Reviewed." *Customs in Common*. E. P. Thompson. London: Merlin, 1991. 259–351.

Thompson, Michael."High on the Hog: Swine as Culture and Commodity in Eastern North Carolina." PhD diss. Miami University of Ohio, 2000.

Tise, Larry E. *The American Counterrevolution: A Retreat from Liberty, 1783–1800*. Mechanicsburg, Pa.: Stackpole, 1998.

Tucker, Robert W. *Empire of Liberty: The Statecraft of Thomas Jefferson*. New York: Oxford University Press, 1990.

Ulmer, Benjamin F. *Lumbermen's Yellow Pine and Cypress Reference Book: A Complete Compendium of All the Rules Governing the Classification and Inspection of Yellow Pine and Cypress Lumber and Crossties, Now Used by the Various Lumber Exchanges and Associations; Also Used by the Principal Railroads of the United States*. Atlanta: Southern States Publisher, 1905.

United States Bureau of Foreign and Domestic Commerce and Forest Service. *Directory of American Sawmills*. Washington, D.C.: Government Printing Office, 1915.

United States Bureau of the Census. *Census of the United States, 1790–1870*. 23 vols. New York: Norman Ross Publishing, 1990.

———. *Tenth Census of the United States, 1880*. 22 vols. New York: Norman Ross Publishing, 1991.

———. *Report on Population of the United States at the Eleventh Census*, 1890. New York: Norman Ross Publishing, 1993.

———. *Twelfth Census of the United States Taken in the Year 1900*. 10 vols. New York: Norman Ross Publishing, 1997.

———. *Thirteenth Census of the United States Taken in the Year 1910*. 11 vols. New York: Norman Ross Publishing, 1999.

———. *Fourteenth Census of the United States Taken in the Year 1920*. 15 vols. New York: Norman Ross Publishing, 2001.

———. *Fifteenth Census of the United States, 1930*. 36 vols. New York: Norman Ross Publishing, 2003.

United States Department of Agriculture. *Silvics of North America, Volume 1: Conifers*. Washington, D.C.: Government Printing Office, 1990.

United States Department of the Interior. *Reclamation and Rural Development in the South*. Washington, D.C.: Government Printing Office, 1927.

United States Forest Service. *Timber Depletion, Lumber Prices, Lumber Exports, and Concentration of Ownership: Report on Senate Resolution 311*. 3rd ed. Washington, D.C.: Government Printing Office, 1928.

United States Forest Service. "Smokey says, burned timber builds no homes: prevent forest fires," govt. poster. Washington, D.C.: GPO, 1946.

United States Forest Service. *A Forest Atlas of the South*. Washington, D.C.: United States Department of Agriculture, 1969.

Vance, Rupert B. *Human Geography of the South*. Chapel Hill: University of North Carolina Press, 1932.

Vickers, Daniel. "Competency and Competition: Economic Culture in Early America." *William and Mary Quarterly* 47, no. 1 (January 1990): 3–29.

Vileisis, Ann. *Discovering the Unknown Landscape: A History of America's Wetlands*. Washington, D.C.: Island Press, 1997.

Wahlenberg, W. G. *Longleaf Pine, Its Use, Ecology, Regeneration, Protection, Growth, and Management*. Washington, D.C.: United States Department of Agriculture, 1946.

Wahlenberg, W. G., S. W. Greene, and H. R. Reed. *Effects of Fire and Cattle Grazing on Longleaf Pine Lands, as Studied at McNeill, Miss*. USDA Technical Bulletin 683. Washington, D.C.: United States Department of Agriculture, 1939.

Walker, Laurence C., and Brian P. Oswald. *The Southern Forest: Geography, Ecology, and Silviculture*. Boca Raton, Fla.: CRC Press, 2000.

Wallerstein, Immanuel. *Historical Capitalism, with Capitalist Civilization*. London: Verso, 1996.

———. *The Modern World-System*. New York: Academic Press, 1974.

———. *The Consolidation of the European World-Economy, 1600–1750*. New York: Academic Press, 1980.

Walsh, Brian J., and J. Richard Middleton. *The Transforming Vision: Shaping a Christian Worldview*. Downers Grove, Ill.: InterVarsity Press, 1984.

Ware, Stewart, Cecil Frost, and Phillip Doerr. "Southern Mixed Hardwood Forest: The Former Longleaf Pine Forest." *Biodiversity of the Southeastern United States: Lowland Terrestrial Communities*. Edited by William H. Martin, Stephen G. Boyce, and Arthur Echternacht. New York: Wiley & Sons, 1993. 447–493.

Watson, Alan D. *Wilmington: Port of North Carolina*. Columbia: University of South Carolina Press, 1992.

Watson, Winslow C., ed. *Men and Times of the Revolution, Or: Memoirs of Elkanah*

Watson, Including His Journals of Travels in Europe and America from the Year 1777 to 1842 and His Correspondence with Public Men, and Reminiscences and Incidents of the American Revolution. 2nd ed. New York: Dana & Company, 1857.

Weakley, Alan S., and S. Kay Scott. *Natural Features Summary and Preserve Design for Carolina Bays in Bladen and Cumberland Counties, North Carolina*. Raleigh: North Carolina Department of Natural Resources, 1982.

Wernicke, O. H. L. *The Lumbermen's Obligation to the South*. New Orleans: Southern Pine Association, 1918.

Wetherington, Mark V. *The New South Comes to Wiregrass Georgia, 1860–1910*. Knoxville: University of Tennessee Press, 1994.

Whedbee, Charles Harry. *Legends of the Outer Banks and Tar Heel Tidewater*. Winston-Salem, N.C.: John F. Blair, 1966.

White, Newman Ivey, et al., eds. *The Frank Brown Collection of North Carolina Folklore: Games and Rhymes, Beliefs and Customs, Riddles, Proverbs, Speech, Tales, and Legends*. Vol. 1. Durham: Duke University Press, 1952.

White, Richard. *The Organic Machine: The Remaking of the Columbia River*. New York: Hill and Wang, 1995.

———. *Land Use, Environment, and Social Change: The Shaping of Island County, Washington*. Seattle: University of Washington Press, 1992.

———. *The Middle Ground: Indians, Empires, and Republics in the Great Lakes Region, 1650–1815*. Cambridge: Cambridge University Press, 1991.

———. "'Are You an Environmentalist or Do You Work for a Living?': Work and Nature." *Uncommon Ground: Rethinking the Human Place in Nature*. Edited by William Cronon. New York: W. W. Norton, 1996. 171–185.

Wiles, Richard, Jacqueline Savitz, and Brian A. Cohen. *Particulate Air Pollution: Human Mortality, Pollution Sources, and the Case for Tougher Clean Air Standards*. Raleigh: Environmental Working Group, 1997.

Williams, Brett. "The Concept of Community," *Reviews in Anthropology* 31, no. 4 (January 2002): 339–350.

Williams, Justin. "English Mercantilism and Carolina Naval Stores, 1750–1776." *Journal of Southern History* 1 (May 1935): 169–185.

Williams, Michael. *Americans and Their Forests: A Historical Geography*. Cambridge: Cambridge University Press, 1988.

Wilton, Andrew, and Tim Barringer. *American Sublime: Landscape Painting in the United States, 1820–1880*. Princeton: Princeton University Press, 2002.

Wood, Bradford. *This Remote Part of the World: Regional Formation in Lower Cape Fear, 1725–1775*. Columbia: University of South Carolina Press, 2004.

Wood, Gordon S. *The Radicalism of the American Revolution*. New York: Vintage, 1993.

———. "The Enemy Is Us: Democratic Capitalism in the Early Republic." *Wages of Independence: Capitalism in the Early American Republic*. Edited by Paul A. Gilje. Madison: Madison House, 1997. 137–153.

Wood, Peter H. "The Changing Population of the Colonial South: An Overview by Race and Region, 1685–1790." *Powhatan's Mantle: Indians in the Colonial Southeast.* Edited by Wood, Gregory A. Waselkov, and M. Thomas Hatley. Lincoln: University of Nebraska Press, 1989. 35–103.

———. *Black Majority: Negroes in Colonial South Carolina from 1670 through the Stono Rebellion.* New York: Norton, 1975.

Woodman, Harold D. "Economic Reconstruction and the Rise of the New South, 1865–1900." *Interpreting Southern History: Historiographical Essays in Honor of Sanford W. Higginbotham.* Edited by John B. Boles and Evelyn Thomas Nolen. Baton Rouge: Louisiana State University Press, 1987. 254–307.

Woodward, C. Vann. *The Origins of the New South, 1877–1913.* Baton Rouge: Louisiana State University Press, 1951.

Wright, Gavin. *The Political Economy of the Cotton South: Households, Markets, and Wealth in the Nineteenth Century.* New York: W. W. Norton, 1978.

———. *Old South, New South: Revolutions in the Southern Economy since the Civil War.* Baton Rouge: Louisiana State University Press, 1986.

Index

Tycho de Boer is assistant professor of history at Saint Mary's University of Minnesota.